BARBERS' MANUAL,
TEXT BOOK ON TAXIDERMY

BARBERS' MANUAL,
TEXT BOOK ON TAXIDERMY

T. J. MCCONNAUGHAY

ISBN: 978-93-90294-37-4

Published: 1898

LECTOR HOUSE LLP
E-MAIL: lectorpublishing@gmail.com

T. J. MCCONNAUGHAY.

BARBERS' MANUAL,
TEXT BOOK ON TAXIDERMY:

BARBERS' MANUAL
(PART FIRST),
TAXIDERMIST'S MANUAL
(PART SECOND)

BY

T. J. MCCONNAUGHAY

1898

PREFACE.

By way of explaining the dual character of this little book, we here indulge a few brief introductory remarks.

Part First is designed simply as a manual of instruction for barbers, and we hope it will, in a valuable measure, supply a long felt need. It will prove especially beneficial to the younger and less experienced members of our craft.

The best artisans and artists admit they owe much to the accumulated knowledge and experience of both their predecessors and their contemporaries. Indeed, to assert any claim to advanced knowledge and skill, without due acknowledgements for the help received from others would savor of an inexcusable egotism. The man who professes to be a self-made man is always notorious for worshiping his maker.

We should, of course, always put our own impress upon all our work. Our observations on the methods of others should supplement but not supplant our own originality and our own reason. A noted artist when asked how he mixed his paints to achieve such wonderful results, replied: "I mix them with brains."

So must we use our own brains as well as the brains of others, if we would succeed in this day of rapid improvements and sharp competition.

The information given in this book is not guess work, but the result of long years of study and practice. Thirty-one years have I conducted a shop of my own. During these years every recipe here given has been thoroughly tested. We know they are all good.

Part Second is devoted to taxidermy, which for twenty-three years I have connected with my other shop work. I have preserved and sold hundreds of specimens, and this work has proven a source of much pleasure and profit, enabling me to turn many otherwise idle moments into money.

On this line I have not confined myself to taxidermic work alone, but have also connected the tanning of hides with the hair on, which I have manufactured into mats, rugs, etc.

Hence this little book, under a twofold title, practically includes three trades.

We offer it as a money saving and money making investment. The recipes and other information contained herein will enable any barber to make all his own preparations, and to manufacture them for sale. He may also connect taxidermy and tanning if it suit his pleasure and business.

CONTENTS

BARBERS' MANUAL
(PART FIRST)

HISTORIC SKETCH.

The word barber is derived from the Latin word "barba," which means beard, and hence is applied to one whose occupation is to shave and trim beards and cut hair. The barber's craft is a very ancient one. The practice of shaving was common among the ancient Egyptians, Greeks and Romans; and was even mentioned by the prophet Ezekiel, chapter v, 1st verse. Among the ancient Israelites the removal of the beard by shaving or plucking was a sign of mourning. It would seem that the origin of our trade was prehistoric.

In early times barbering was conjoined with surgery, and the barber also served the public in the capacity of surgeon. In France the barber-chirurgeons, or barber-surgeons, were separated from the perruquiers, or wig-makers, and were incorporated as a distinct body in the latter part of the 17th century. In England barbers were incorporated with special privileges in 1461, and were afterwards united to the society of surgeons, but were confined to the minor operations of blood letting and drawing teeth. In 1745 an act was passed, the preamble of which declared the trade of the barber and the profession of surgery are foreign to and independent of each other.

This act practically divorced surgery from the barber's chair. However, quite recently the surgeons of the Swedish navy were also barbers for the crew. In former times, not far remote, it was customary to keep a lute or a violin for the entertainment of patrons, which made the shop a favorite resort for idle persons. In China and some other oriental countries, barbers not only shave the face, but they also shave the whole or a part of the head.

BARBER'S SIGN.

Though barbering is now entirely separated from surgery, yet a monument of their former union may be still seen in the striped pole and basin so often projecting from the front of barber shops.

The fillet, or spiral strip around the pole symbolizes the winding of a ribbon round the arm previous to letting blood, and its red color symbolizes the blood. The basin, yet sometimes seen at the base, represents a vessel prepared to catch the blood.

Barber's poles in this country are tri-colored. The white symbolizes the arm, the red represents the shed blood, while Americans have added the blue to complete our national colors.

RECIPES, FORMULAS, ETC.

GERMAN HAIR DYE.

The process here given comprehends a double formula, and to insure satisfactory results, the directions subjoined should be carefully followed.

No. 1.

Nitrate of Silver	3 drachms
Stronger Water of Ammonia	½ ounce
Distilled or Rain Water	½ ounce

No 2.

Pyrogallic Acid	3 drachms
Alcohol	½ ounce
Distilled or Rain Water	6 ounces

Directions.—Formula No. 1, and formula No. 2, should each be put up in a separate bottle.

In compounding, each of the first two ingredients should be combined and reduced to a solution and the water then added.

Before using these preparations the beard or hair should first be thoroughly washed and dried.

Then apply No. 1 with the fingers, and when partly dried apply No. 2 with a tooth brush and a comb, being careful to get the dye down to the skin. If convenient let the party go out into the open air for an hour. Otherwise the hair should be thoroughly dried with a fan, and washed with soap and soft water. This formula was brought from Germany by a St. Louis barber and sold to the author in 1869. Properly put up it is not only first-class, but I have never found any other preparation equal to it.

BROWN HAIR DYE.

Take four pounds of green walnut hulls. Put them in 1½ gallons soft water and boil down to three pints. Strain off through cloth until clear of sediment. To one quart of this add one quart of alcohol, and 3 ounces of glycerine. Use as other restorers, once a day until the desired shade is obtained. The hulls should be gathered in August.

QUININE HAIR TONIC AND SEA FOAM.

The strong point in favor of this preparation as a sea-foam is that it acts at the same time as a tonic for the hair.

FORMULA.

Alcohol	1 pint
Glycerine	½ ounce
Tincture of Cantharides	½ ounce
Aqua Ammonia	½ ounce
Sulphate of Quinine	30 grains
Oil of Cloves	1 drachm
Rock salt (or table salt)	¼ ounce
Distilled or rain water	1 pint

Directions.—When using as a sea-foam, wet the hair and rub briskly with finger ends until the foam has disappeared. Then take a towel and rub partly dry.

When using as a dandruff cure, sea-foam twice a week, dampen the hair twice a day for two weeks, after which use once a week as a sea-foam. This will keep the scalp in fine condition.

After you have tried it you will use no other.

CREAM FOR CHAPPED HANDS AND FACE.

This recipe was given me by a druggist whom I have known for years. Its reliable and competent source is a sufficient guarantee:

FORMULA.

Quince seed	¼ ounce
Distilled extract witch hazel	1 pint
Glycerine	¼ pint
Alcohol	¼ pint
Powdered Boracic Acid	160 grains
Carbolic Acid	32 drops
Perfume to suit.	

Directions for Making.—First put quince seed in witch hazel and let stand twenty-four hours. Then strain through cheese cloth and add the other ingredients. I know this to be good for the face after shaving.

PYTHIAN CREAM.

This is another preparation for the face and hands.

FORMULA.

Gum Tragacanth (in flake)	½ ounce
Glycerine	4 ounces
Distilled or rain water	½ gallon

Directions for Making.—Put gum tragacanth in water and let stand till thoroughly dissolved, and strain through a cheese cloth. Then add the glycerine and a sufficient amount of Pythian bouquet to perfume. Color pink with powdered carmine. It should be about the consistency of cream. If too thick add more water. There should be a half-gallon water to the four ounces of glycerine when finished.

BAY CREAM.

The only difference between this and Pythian Cream is in the perfume. Some like the bay rum better.

Directions.—Use the formula of the preceding and then add perfume with the genuine oil of bay.

COLD CREAM.

The only difference between this and the two preceding creams is that instead of perfume a compound of alcohol and menthol is used.

Directions.—Use the same body as for Pythian Cream, and then add alcohol and menthol as follows:

Put 6 drachms of menthol crystals into an ounce of alcohol. When the menthol is thoroughly dissolved add this combination to ½ gallon of the cream. The presence of the menthol gives this preparation a very pleasant cooling effect. Hence it is well named. This cream may, of course, be perfumed if desirable.

Instead of these face creams, some customers will prefer the pure bay rum, while perhaps many will prefer witch hazel. These can, of course, be obtained from barber supply houses or from drug stores; the witch hazel may be improved by adding to it a good cologne. Try two parts witch hazel to one part cologne.

BAY RUM.

No. 1.

Bay Oil	½ ounce
Oil of Pimento	¼ ounce
Alcohol	3 pints
Water	3 pints

No. 2.

| Magnesium | ¼ pound |
| Oil of Bay | ¼ pound |

Mash them well together and put them in a filter and pour in two quarts of water. Let it filter slowly, and then add 2 quarts Alcohol.

IMITATION OF BAY RUM.

No. 3.

Oil of Bay	3 drachms
Oil Pimento	½ drachm
Water	1½ quarts
Acetic Ether	1½ ounces
Alcohol	2 quarts

Mix and let stand 3 days, then filter.

CAMPHOR ICE.

Oil of Sweet Almonds	2½ ounces
White Wax	2½ ounces
Spermaceti	3½ ounces
Gum Camphor	¾ ounces

Mix together, melt and pour off in small salve boxes.

MENTHOL SALVE.

Mutton Tallow	1 ounce
Lard	1 ounce
Menthol (in crystals)	3 drachms

Melt together and pour off in salve boxes.

Both the Camphor Ice and the Menthol salve are good for tender faces.

SILVER GLOSS SHAMPOO.

This is an economic and very satisfactory preparation.

FORMULA.

White Castile Soap (the very best)	1 pound
Refined Carbonate of Potash	¾ pound
Distilled or rain water	1 gallon
Table Salt	½ ounce

Refined Carbonate of Potash is also called Pure Salts of Tartar. I have found the English brands preferable.

Directions.—Shave the soap fine and put into the water (as per above formula), which should be contained in a porcelain vessel.

Let it boil until soap is thoroughly dissolved and strain off into another vessel, and then add the pure salts of tartar while still hot. Add the salt and enough more boiling water to replace the amount which has boiled away, and continue to stir until it becomes only luke warm; then add a few drops of the oil of cloves (or some other perfume), if desirable. Finally pour off in small jelly jars and set away for use. 1 gallon made in this way will make 5 gallons of ordinary shampoo, by simply adding 4 gallons more of water. This quantity should not cost over fifty cents. A pound of the refined carbonate of potash costs twenty cents, and a pound of castile soap only fifteen cents, and the perfume will cost less than the remainder of 50 cents.

One teaspoonful is enough to clean any ordinary suit of hair.

In cleaning ladies' hair it is well to add a little ethylic ether, commonly called sulphuric ether. Never use hard water. If necessary save up enough rain water. I give elsewhere directions for making shampooing outfit, which may also be conveniently used for shower baths in shops and houses where there is no connection with water works. Water to be used for shampooing should always be warm.

EGG SHAMPOO.

This favorite preparation should be used immediately after mixing.

Take 1 fresh egg, 1 teaspoonful of silver gloss shampoo, and ¼ teaspoonful of powdered borax. Mix together with an egg beater, and then use as other shampoos.

A CHEAP SEA FOAM.

Take 2 ounces of the silver gloss shampoo, 2 ounces alcohol, 1 ounce glycerine and 1 pint water; shake well together and perfume to suit your fancy.

The shampoo or sea-foam can be colored a nice yellow by making a tea of saffron and water, adding enough after straining it to get the desired color. Powdered carmine can also be used to color a red or pink color.

BRILLIANTINE.

Take 1 ounce of good glycerine, ¼ ounce of rose geranium and 1 ounce water. Mix. This preparation is a good one, and can be made very cheap by using a less amount of the perfume. It never separates, and is good as long as there is a drop of it left.

ENGRAVING FLUIDS.

We here give a formula for making an etching fluid, to be used in marking razors, shears and other steel tools.

FORMULA.

Bluestone	1 ounce
Table Salt	1 ounce
Water	6 ounces

Directions.—Cover blade or plate with soap or varnish, and then with etching needle or common pencil write the name or letters desired, being careful to score or scratch through to the metal. Then fill the traced lines with the fluid and let it remain five minutes. The fluid will corrode the metal in the lines thus laid bare. Therefore when the covering and acid are washed off the lettering will remain.

Be careful to wash promptly and dry thoroughly.

BLACK HEADS.

What are known as black heads are generally found in the skin of people who are addicted to the use of much hog meat. Such people are also as a rule, rather careless, to say the least, about bathing their faces. A hint to the wise will be sufficient. Let them not be afraid a rough towel will scratch them. I give below a recipe highly recommended.

FORMULA.

Alcohol	4 ounces
Boracic Acid	2 drachms
Distilled or rain water	1½ ounces

Apply this three times per day after first having thoroughly washed the face and rubbing dry with a coarse towel. Considerable benefit will, at least, be derived from a faithful application of the above.

HAIR BLEACHING.

First clean the hair with the Silver Gloss shampoo, and when dry apply peroxide of hydrogen until damp. When dry, again repeat the application, and continue to repeat it until nearly as light as desired.

The hair will continue to bleach a little lighter for about three days, and hence it is necessary to discontinue the application when the hair is a shade darker than desired.

WHITENING FOR THE FACE.

Put 1 ounce of the oxide of zinc into a plate and pour over it 3 ounces of soft water. Mash zinc with a spoon until it is all dissolved. Pour the solution into a pint bottle and fill up with witch hazel. When the weather is cold, pure soft water may be used instead of the witch hazel; but the preparation would sour in warm weather. Apply with a soft cloth.

BARBERS' ITCH.

Fear of this disease causes many men to shave themselves, and this class would otherwise be among the very best customers.

When these men observe how careless the average barber is with his towels, mugs, tools, etc., they become disgusted and purchase a shaving outfit and quit the barbers' chair, except when they want a hair-cut.

I believe every barber should know how to treat this disease. Hence I will make a few suggestions as to its causation and treatment.

Scabies, or itch, in its various forms is a disease caused by the irritation produced from the presence in the skin of what is called the itch mite and the ova of the same. The cure involves the destruction of these parasites. Get a doctor, if possible, to prescribe; if no doctor can be got who understands it, I would try the following: An ointment made from the flour of sulphur and lard or sulphur and vaseline, is about the best remedy known. Rub in well at night and wash off in morning. Or take citron ointment 1 ounce and mutton tallow 1 ounce. Melt together and stir till cool. This I have found one of the best salves for all skin diseases I have ever tried. Apply twice a day, but use with care since it contains mercury.

HAIR OILS AND HAIR DRESSINGS.

While oiling the hair is a thing of the past, we might indulge a few remarks as a matter of history. Thirty years ago almost every customer used oil on his hair, and every barber was expected to know how to mix his own oils. A favorite preparation was made as follows: 1 pint of alcohol and 1½ pints of castor-oil were shaken together, and then perfumed with citronella or bergamot.

Another favorite was made of raccoon oil and lard mixed half and half, and perfumed with the oil of cloves.

Some used the coon oil straight; others used the oil of birds, geese, chickens or ducks, etc. Bear oil was considered a great oil for the hair as well as for many other purposes. My own favorite among all the home made preparations was made from beef marrow. The marrow was tried out and a little salt was added. The oil was then perfumed with bergamot.

POMADES.

In selecting material for pomade, have a butcher get you some fine leaf lard and some of the finest suet, which should be taken from young animals. Render out separately in porcelain vessels and strain off.

Directions.—Take lard 1 pound, tallow 1 pound; mix them and heat gently, and cook for one hour over a slow fire; remove and let stand a few minutes to settle; now pour off carefully. When almost cold add some suitable perfume, say oil of bergamot 4 drachms, oil of lemon 3 drachms, oil of cassia 2 drachms, oil of nutmeg 75 drops. Mix thoroughly with the pomade and pour into small jars.

STICK POMADES.

Take of the prepared tallow 1 pound, pure, clean bees wax 3 ounces, gum benzoin, in a coarse powder, 1¼ drachms. Melt together with a slow heat, mixing all the while. When partly cooled add some suitable perfume. Pour it off in moulds and when cold take out and wrap in tin foil, then put on a nice label as outer covering.

To make the above into a coloring pomade, take 3 pounds of the prepared lard and tallow, before being perfumed; add to it 2 pounds of fresh walnut hulls, cut up fine; put into a porcelain vessel and heat gently for four hours. Take off and strain, and proceed as in making the black pomade. This will gradually color the hair or beard to a nice brown by being used daily until the desired shade is obtained.

HOW TO STOP BLOOD.

Every barber should have at hand a preparation for stopping blood. The best of barbers are liable to bring blood from rough or tender faces. An astringent pencil, which is very good and very handy, may be obtained from the barber supply houses in the cities at a cost of only 10 cents each. However, I prefer Monsell's Powdered Iron which may be obtained from any drug store. The only objection to it is, it is liable to discolor the skin. However, by being careful to put on only a small amount, it may easily be washed off when the blood has ceased to ooze.

If a small bump has been cut off or a shallow cut made in the smooth skin, it will generally suffice to cover it with a thick lather and let it remain until the shaving is completed.

Alum is also used but is too slow in its action.

HAIR RESTORERS.

I could give a formula that would make the hair fall out, but thus far I have not been able to find a preparation that will produce a new growth of hair on bald heads. As a preventative treatment I might suggest as follows:

Boil burdock root in soft water until strong, and then add to one pint of it, a half pint of alcohol, a teaspoonful of salt, and 1 ounce of glycerine. This used once a day will prevent the hair from falling out. Or make a strong decoction of black tea or sage and mix with the alcohol, salt, and glycerine as above and use as above.

The basis of most hair tonics is the tincture of cantharides, quinine, ammonia, camphor, and salt. A solution of borax in camphor water is used by some as a stimulant for the scalp. I have a friend who is experimenting on a new line with very encouraging prospects of successfully producing hair on bald heads. If any party interested will address me a few months hence, I may be able to advise them how to reproduce hair on bald heads.

GLASS HONES.

A glass hone is easily made, and no barber should be without one. Procure a piece of heavy plate glass and have a glass cutter cut it into pieces 3×8 inches in dimensions. Take the gloss off the face and also around the edges on a grindstone, and then finish by rubbing the face of the hone with pumice-stone kept wet with water. Continue this rubbing until the gloss is entirely removed and the hone is smooth. Before honing take a rubber, such as is used on a water hone, and, after wetting hone rub until you have a sort of lather. Hone on this as you would on any other hone. You will find it excellent for smoothing shears after grinding, or a razor after having been over honed.

MUGS, BRUSHES AND SOAPS.

The mug should be large and heavy and the brush used to make the lather should also be large and first-class in every particular. In regard to brushes, I would suggest that it pays to buy the very best.

A poor brush that is continually shedding hairs is very annoying to the customer, and it hinders the barber. I prefer the rubber ferruled brush, but be sure to get the genuine.

In regard to soap I must admit that I am partial to the J. B. Williams barber soap. However, there are other brands that give good satisfaction. There are no soaps too good. Hence get the best.

FACE POWDERS.

There is nothing much better than cake magnesia, but it should only be used to dry the face after shaving. I have given a liquid whiting which is much used by ladies. I have often used it on men to whiten the skin. See whitening for the face.

SHOWER BATH AND SHAMPOO CAN.

Take a common tin bucket which holds three gallons, have a small tube one inch long and one quarter of an inch in diameter put in one side about one half inch from bottom of bucket.

Then get a rubber hose three feet long of suitable diameter to fit on to the tube. At the other extremity of the hose attach a sprinkler with its tubular end made to fit the hose.

Connect the hose to bucket and the sprinkler to hose, and the can is complete.

Lay the sprinkler over the upper rim of can (or bucket) to keep the water from flowing out. Fill with soft warm water; and, when ready to take the bath or shampoo, hang from ceiling or set on shelf high enough to allow the water to flow over the head and body.

Put on the shampoo. Take the sprinkler and thoroughly wash and rinse.

Three gallons of water used in this way is better than a whole bath tub full used in the ordinary way. This apparatus of course applies to country places where they have not the advantage of water works, or where the water furnished is hard.

TOWELS, HAIR CLOTHS, AND FACE CLOTHS.

If you would secure first-class trade, you must keep a good supply of clean good looking towels. Nothing is more disgusting in a shop than a lot of dirty ragged towels.

The workmen may be first-class, the tools first-class, etc.; but unless the towels are in proper condition the better class of trade will go elsewhere.

I prefer a good moderate sized cotton towel, except for the wash cloth, where I prefer a cotton crash towel on the order of a bath towel.

This crash comes in bolts and may be cut the desired length; it must, however, be hemmed. For the bath always use a good towel but not too large.

The breast cloth made of calico or gingham, should be full width of goods, and not less than three and a half feet long. It should have a half neck opening at one side made to fit up around the neck.

The hair cloth may be made of calico, gingham or bleached sheeting. If the sheeting is used get the goods wide enough to require no seam, and then border with a two inch band of red oiled calico.

SUGGESTIONS FOR BEGINNERS.

In the first place, if you have made up your mind to be a barber, why not be a good one. There is always room at the top, and rich reward for him who has reached the top. Rich reward, however, is the price of self-exertion. Do not wait for a tidal wave to waft you on to success. The minions of fortune are few and far between. You must not only work but you must embrace every opportunity to improve your qualifications, if you would achieve success in this age of advanced knowledge and skill. You should always be on the alert, and never miss an opportunity to acquire useful information. Knowledge is power, and it behooves you to gain all the knowledge you can, especially of your own business. If necessary pay for it, work for it, or even beg for it. The possession of a good fund of practical knowledge with other necessary qualities of mind and heart, will enable one to wear good clothes, make plenty of money and to have scores of friends; while the lack of it will make another the digger of ditches, living in rags and poverty, and deprived of the more congenial companionship of the better and nobler elements of society. Knowledge pays.

Moreover, while you should attend well to the proper equipment of your mind, you should not neglect the proper cultivation of your social character. Especially should you cultivate what might be termed a practical business social tact.

Treat every customer as though your success depended upon him and him alone, and always exert your utmost to do good work. Much depends upon the barber's ability to shave well. To accomplish the great desideratum of being able to shave well each one of the various customers with their varying qualities of beard, you must diligently study the temper of your razors with reference to the peculiar beard of each customer. Again much depends upon the lather, the brush, the hone, the strop, etc. Hence you will please pardon a few simple primary suggestions leading up to a good shave. First you must equip yourself with first-class tools.

Procure a first-class oil hone.

A large rubber ferruled lather brush.

A large heavy shaving mug.

A half dozen No. 1 razors, 4½ to 4⅝ wide and ¾ concave.

Your soap and all other materials and implements should be as good as you are able to buy.

When you hone your razor on the oil hone, use good soap and make a stiff lather which you will spread on the hone. Then place the razor on the hone and draw very lightly from heel to point so that the edge of the razor will always be on the front side of the moving blade. Each stroke across the hone should be a sloping or sawing stroke, and at the end of each the razor should be turned on the back and pushed up across the end of the hone, and the other side of the blade laid flat on the hone. Then draw the razor back with a curving or angular stroke to the other end of the hone. Turn again as before and continue with steady stroke until you think the blade is sharp. Wet the thumb nail and try the edge by drawing it lightly from end to end. Should the edge feel blunt or rough make a few more strokes on the hone and try again, and so on until the edge is satisfactory. A better way to test the edge, if you have a smooth soft hand, is to wet the end of thumb or fore finger and draw the razor lightly over it feeling of the edge. However, it requires practice to tell with certainty by either method. Therefore practice carefully until you become expert.

We will here make a few suggestions in regard to the different hones generally used by barbers. The oil hone, which is supposed to be petrified hickory, is the one most used, and is perhaps the best. Second in rank and general use comes the Swatty hone which is made from the same material as the emery wheel, but it is of finer grade. This is a very fast cutting hone. Next comes the water hone which is simply a fine grit stone, and then comes the glass hone which is but little used. A few barbers use it to take off the wire edge of over-honed razors. Hones should be handled with great care. In honing the razor should be run well out to the ends to prevent hollowing the hone. Should a hone begin to hollow, work it down with fine sand paper to a perfect face and smooth it with the rubber used on the water hones. I prefer olive oil on an oil hone if carefully used. Of course lather may be used on an oil hone the same as on the Swatty and glass hones. Use water on the water hone and rub with the rub stone commonly used until the surface is covered with a sort of soapy pasty lather before honing. The glass hone may be used in the same manner. Any hone when not in use should be wrapped up and laid away carefully after having been washed and thoroughly cleaned.

You should be provided with a good shell strop, or a good Russian leather strop, and also a good canvas strop. After honing strop the razor lightly on the leather only. The canvas should not be used except when the razor has become smooth. When you have a customer in the chair, first put a clean towel on him and proceed to make the lather using warm soft water. Lather the face and scour the beard, and then lather again with a good heavy lather. Proceed to shave, drawing the razor with a sawing stroke, and make the strokes as long as practicable. Hold the skin tight to throw out the beard, and go over the face the first time as quickly as possible. Then wash the soap from the face, and go over it a second time, keeping the skin somewhat stretched, and wet with soft water. When finished, press a hot towel to the face and then use one of the face creams given in this book.

Next powder the face, and curl and perfume the mustache. If you think his hair needs trimming tell him so, and if he has it done, do your very best, even taking pains to cut the hair out of his ears. Shave his neck. The most important point is to get a good edge on the hair.

If he takes a shampoo, first give the scalp a good brushing to loosen the dandruff, and then use Silver Gloss Shampoo, giving him to understand that you have it for sale for family use. Rinse hair with warm soft water, and dry with a fine bath towel, and then ask him if you shall apply some of the hair tonic which will cost him only ten cents extra.

Comb his hair in the latest style, and if he desires it, color his mustache with the celebrated German Hair Dye. Help him into his coat, and thanking him bid him come again.

Purchase every good book pertaining to your business, study diligently and practice what you learn, and you will soon stand abreast with the best and most progressive barbers. You should be prompted by no meaner ambition.

SUCCESS.

"If you wish success in life,
Make perseverance your bosom friend,
Experience your wise counselor,
Caution your elder brother,
And hope your guardian genius."

Addison.

TAXIDERMIST'S MANUAL (PART SECOND):
TEXT BOOK ON TAXIDERMY

BY

T. J. MCCONNAUGHAY.

TAXIDERMY.

The word taxidermy is derived from the two Greek words, taxis, which means arrangement, and derma which means skin. Hence this term is applied to the art of preserving and mounting the skins of animals for ornamental and scientific purposes. Little is known of the origin of this art, but it would seem from books of travel and natural history, that it is at most, not more than three hundred years old. It began to be practiced in England about the beginning of the 18th century, which fact is proven from the "Sloane Collection" which was formed in 1825, as the nucleus of the present natural history collection lodged in the galleries of South Kensington. It was about the middle of the 18th century that the first book devoted to the principles of taxidermy was published in France. After this, others appeared from time to time in France and Germany, but England contributed no literature on the subject until about the beginning of the present century. In 1828 an English-man named Scudder, established a museum of mounted specimens in an old alms house in New York City. Previous to this, the art seems to have been absolutely unknown in America. It was not till the exhibition of 1851, that the French and German taxidermists taught the English the principles of scientific treatment.

Since that time several works have appeared from the pens of English and American authors. Prominently among the American writers, were Charles Waterton and Titian R. Peale who greatly improved the art in this country.

Jules Verreaux, of Paris, brought the art to a still higher perfection, and introduced methods for giving to specimens a life-like expression, which elevated it quite to the realm of higher art. Great were his accomplishments in the art of expressing the actions and characteristic attitudes of the living animals. Since his day taxidermy has rivaled the plastic art, and today, it has reached such a degree of perfection that the most artistic and æsthetic effects may be wrought by the hands of any operator who possesses artistic faculties. The Illinois State Natural History Society of Bloomington, published an illustrated pamphlet from the pen of one Mr. Holder, which is a very valuable contribution to taxidermic literature. The author was doubtless greatly improved by his associations with Audubon and Bell, and in his book he gives the results of a ripe experience. This book ranks as one of the best yet published. We submit the results of our own patient study and practical experience, and hope it may prove a worthy addition.

ARTICLE I.

ON SKINNING, STUFFING AND PRESERVATION OF BIRDS.

SKINNING.

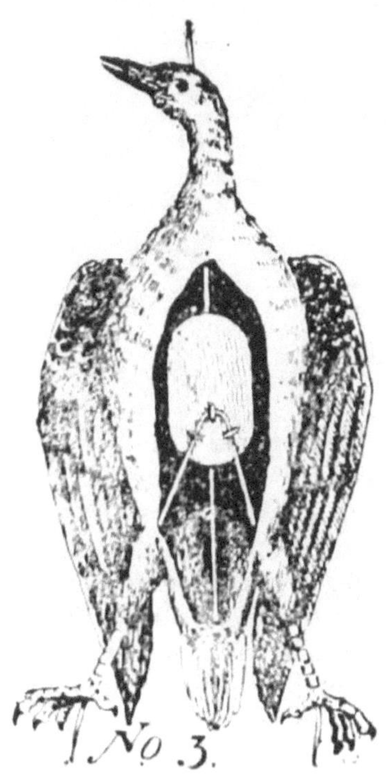

NO. 3.

Immediately after the bird is killed, the nostrils, throat and wounds should be stuffed with cotton to prevent the blood from oozing out and staining the plumage; but should any blood get on the plumage, it should be removed as soon as possible. This can be done by taking a cloth or sponge and dipping it in clean water, wringing it out so as to leave it only moist, and rubbing the feathers gently until all traces of blood stain are removed. Now sprinkle the feathers with plaster of paris, and shake out before it sets. Repeat this until they are perfectly dry, and then wrap the specimen in paper to protect the plumage until ready to skin, which should be done as soon as the bird cools.

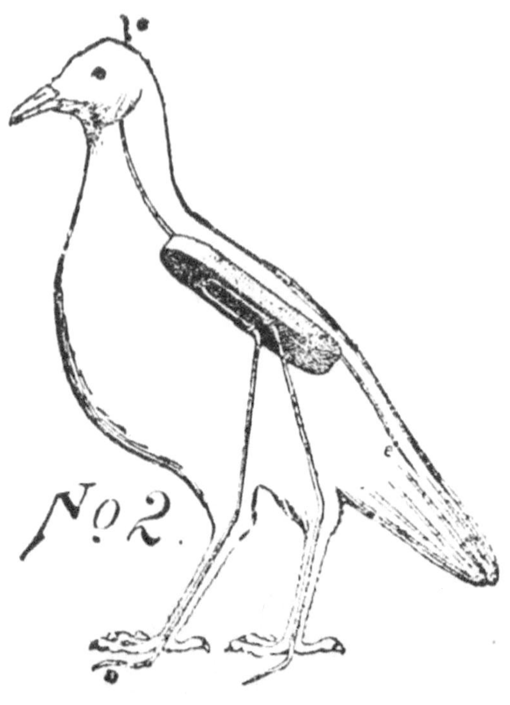

NO. 2.

In proceeding to skin a bird, spread a cloth on a board or table and lay the bird on its back. Separate the feathers on the breast with a scalpel or knife. Insert the knife at the top of breast-bone and cut the skin from there to the tail. Great care should be taken not to cut too deep over the intestines, as it is only necessary to cut through the skin. Now begin where you first inserted the knife and proceed to separate the skin from the flesh, either with the fingers or the back of the scalpel or knife. Tear some small pieces of paper, say about an inch square, and put under the skin on the flesh as you get them separated, which will keep the feathers from sticking to the flesh and becoming soiled; or sprinkling cornmeal over the flesh and skin while skinning, will answer the same purpose. Press carefully down each side to backbone. Now press the thighs forward and inward, draw the skin from the thighs, unjointing them at the first joint from the body.

The skin is now removed over the rump, and the tail unjointed, taking care not to injure the tail feathers. Now for convenience, take a cord and fasten it to a nail on the wall or ceiling, put a wire hook on the other end, and fasten in the bird so you can suspend it high enough above the table to work on it easily. Now pull the skin downwards until you get to the wings, which are to be unjointed at the shoulder joint. It is then pulled down over the skull until the ears are reached. Here many valuable specimens are spoiled by cutting the ears too close to the skin, so be sure to cut the ears close to the skull.

The eyelids are often spoiled also by inexperienced hands. Be sure and cut them well back and, if necessary, trim them afterwards. Now remove the eyeballs and unjoint the skull from the neck, enlarge the opening at the base of skull, where the neck came off, and remove the brain with spoon made for that purpose. Now remove the flesh from the thigh bone down to knee joint, and turn them back in place; then skin the wings out to first joint and remove the flesh. There is yet one job and the skinning process is completed. Divide the feathers on the under side of the wing between the second and third joints, cut the skin, and with a sharp knife cut out all flesh from the bone, and put in some of the preserving powder; cover it with a little cotton and sew it up, being careful not to draw any of the feathers in; press the feathers down smooth, and see that each is in its proper place.

There are some birds with large heads and small necks. With these we cannot draw the skin over the head, but have to push the skin as near the head as it is possible, and cut off the neck bone; then make an opening under the throat large enough to turn out the skull and the remaining part of the neck bone, and proceed to skin the skull, take out brain, etc., as directed heretofore. The flesh must be carefully removed from all parts of the skin, and the preserving powder applied, being very careful that all parts are well powdered.

If you now wish to mount the bird it can be done immediately.

STUFFING BIRDS.

In the first place take some plaster paris, mix it with water to a stiff paste, and fill the eye-sockets with it, then press the glass eye in the paris, using great care to set them in as near a natural position as possible. When the plaster has hardened, which it does very quickly, if the skin has become dry, dampen it with a damp sponge, and turn it back over the skull, then lay the bird on the table and proceed to put in the frame, stuff and sew up.

I give three kinds of frames. One is the wire and wood frame, which you see on page 78, Figs. 9 and 10. The others, wire frames, on page 80, Fig. 3, for birds; also Fig. 11, for humming birds and other small birds, on page 78.

Take frame described Nos. 9 and 10, having the leg wire AA detached from body board, fill the neck with tow, being careful not to fill it too full; run neck wire E through the center of the neck and pass it out at the top of head, as shown in Figs. 2 and 3, or through one of the nostrils. I prefer the latter, because it does not break the skin and holds the head more steady. Then put the tail support E through the center or heavy part of the tail, raise the body board and place under it some tow or cotton; or pad the board by placing some cotton on it and tacking some cloth over it to keep the frame off from the back; then take the leg wires, Fig. 13, AA and put them in the legs, as shown in cut No. 3, and fasten on to the body board with blind staples, as represented in cut No. 3; proceed with the stuffing; finish filling out the neck and breast, shaping it while filling, using care not to get it too full, as that is a fault of most beginners; after filling it down to the opening, the next should be the legs; if the leg bone is left in from the knee up, wrap it to the leg wire with a strip of muslin and tie it; then stuff the leg up to the body, fill the body and sew up; in sewing be careful not to draw any of the feathers down with the thread. After it is all sewed up, place all the feathers down smooth and in their proper place, as much depends on this.

Next place the specimen on a board or perch; if on a board, procure a suitable one, bore two holes the size of the leg wires through it, then make a groove on the under side of board running off from each hole to lay wire down in; place the specimen on the board, running the leg wires through it, and draw the wires down until the legs set in proper position, bend the wires down into the grooves and fasten them with small blind staples. In this connection one should use his own judgment in placing the bird on the board; one leg should generally be placed a little in advance of the other. If the bird is to be placed on a limb perch, bore the holes through the limb, place the bird on, and draw the wires tight, and drive a wooden wedge in beside the wire to hold it, then file the wire off close to the under side of the perch. Pose the bird in as natural a position as you can, imitating nature as near as possible. Place the wings in position, then take a piece of wire heavy enough to hold the wings in place. Sharpen one end, and make a square turn about one-fourth of an inch from the other end to keep it from pulling through the wing. Now place the wings in proper shape. Pass the wire through the wings and body in a way to hold them in shape. Pull the wire until the turn on the one end presses against the wing. Cut the wire on other side of bird and turn it back, as on the other side, so that the wire cannot be pulled either way; then arrange the whole body—wings, neck, tail, head, etc. In case the specimen is a web-footed bird, take a thin piece of board, say from a cigar box, cut it to fit in between the toes, and tack them in to hold the web in shape while drying. This should be removed when dry. After all is completed bathe the buts of wings, the feet, legs and beak, with the preserving fluid; this should be repeated for three or four days. Then let the specimen dry in a shady place. It can then be placed on another board or perch and set in the cabinet.

MOUNTING BIRDS ON WIRE FRAME.

(See directions for making frame.)

Skin the bird and prepare it as directed in this article. Fill the neck with tow, put in neck wire, letting the end come out through the nostril; bend leg wires AA back so as to get them in the legs; run them down on the inside of the leg, or through the center of leg bone, and come out in the center of the foot. (See cuts Nos. 2 and 3.) Put tail support E through the under part of the tail, and proceed as directed heretofore in this article, to stuff, sew up, etc.

Where the tow can not be secured and you have to use cotton in stuffing, always put in the neck wire first, and stuff around it or wrap the neck wire with strips of muslin, old calico, or anything that can be used for that purpose, always being careful not to make the neck too large. Excelsior can be used to good advantage in stuffing large bodies.

PELICANS.

In dressing a pelican always use the board and wire frame. For position, copy after some picture of the bird, which you can find in any natural history. The only difference in dressing from the goose, etc., is they have a game sack under the lower jaw, which is often as much as ten inches wide and sixteen inches long. I find the best way to dress that is to take a fine shingle, wide and long enough, and trim it in the shape of a sleigh runner, and put it inside of the pouch or game sack, with the straight edge up and the square end toward the neck. Draw the pouch smooth over the shingle, and tie the bill together, then wet well with the solution of corrosive sublimate. The board should be left in. Wire can be bent and placed in to answer the same purpose.

PEA-FOWL.

In dressing a pea-fowl, where the wings and tail have to be spread, extra wires have to be put in. The wires for the wings are fastened on to the body board, as when used for the fore legs of animals. Then, for the tail support, take a wire about five feet long and bend it in a hoop shape, leaving enough of the ends to extend into and fasten on the body board. This must be put on the board before putting it in the body. Dress the bird as others, letting the hoop or tail support extend out under the tail. After it is all stuffed and set on the board, bend the tail support up back of tail, and fasten the feathers of the tail to it by tying them, one at a time, with a heavy thread, in such a way that when all are fastened to the wire they will stand as when the bird was strutting. Arrange the wings and body to suit and let dry.

DRESSING BIRDS WITH WINGS SPREAD.

To dress a bird with wings spread, either flying or sitting, it will be necessary to put in wing wires to hold the wings out. The easiest way is to use the wire and board frame, using the front leg wires as used in animals, for the wings. See Figs. 7 and 8, board and body frame.

In dressing a bird this way, after skinning and wiring, as directed for bird, using frames 9 and 10, put in the wing wires and fasten them to body board, as directed for forelegs of animals.

If wishing to represent the bird as flying, take four cords and fasten them to the back of the bird in a way that will balance it; tie them together a few inches above the back and extend one cord longer than the others. It can now be hung from the ceiling by the cord; this will let the specimen turn about and present a fine appearance. The feet should be placed in as near a natural position as possible as when flying. In all large footed birds use plenty of the solution of corrosive sublimate or turpentine. The carbolic acid and water mixture is also used, and is good.

TO CLEAN FEATHERS.

As a test case take a bird after killing and smear it all over with blood, let it lie until dry. Then put it in a basin, and give it a good washing with soap and water, then rinse it clean. Now take a dry cloth and wipe it until almost dry. Then sprinkle it with dry plaster of paris and shake it out before it has time to set. Repeat this powdering and shaking until it is perfectly dry, and the feathers will look bright and nice. Try it.

PRESERVATION OF BUGS, ETC.

Take a large moth fly or miller full of eggs. Make an incision on the under side of body and take out all the eggs and entrails. Now put in some of the preserving fluid and fill the cavity with cotton and sew it up. Then run a pin down through the body and pin it to a board. Arrange its feet, moisten its body with the preserving fluid and the job is complete. Butterflies and small bugs need only the preparation over the body. Heavy bugs can be placed in a cup and the preparation poured over them. Let them lie in it a few hours and pin on board as directed in moth flies. When dry place in a glass case.

ARTICLE II.

OF SKINNING ANIMALS.

As soon as an animal is killed, take cotton and stuff in the wounds, nostrils and mouth, as directed in Article I. for birds; then let it cool before commencing to skin. In skinning a specimen be very careful about cutting holes in the skin, and also not to stretch it by pulling on it while skinning. All being ready, lay the specimen on its back, make an incision from the breast-bone back to the center between the hind legs, being careful not to cut too deep over the abdomen. Divide the skin right and left, putting cotton or paper on the body as the skin is removed, or sprinkle with corn meal, as directed for birds in Article I, to prevent the fatty matter from soiling the hair. When you come to the hind legs, pull the skin down over the thigh joint, and unjoint it, cut the leg off and skin down to the knee joint and unjoint there, taking out all the flesh down to the foot, then treat the other hind leg the same way. Now skin down over the rump and back to the tail. The tail is hard to skin. Take a forked stick, and after pressing the skin as far down on the tail-bone as possible, holding the bone as tight in the fork as you can, pull downward and the bone will slip out. If the animal is not too large it should be swung up with a rope or cord tied around the body just in front of the hind legs, or by attaching a hook to the rope which fastens into the flesh. Remove the skin down to the forelegs, and unjoint them at the body, and skin out in the same manner as directed for the hind legs, unjointing at the knees. Now skin the neck and head, and when you come to the ears cut them close to the skull; fine specimens are often spoiled by cutting them too far out from the skull. We next come to the eyes. Be careful not to cut the eyelids. It is better to cut close to the skull and trim afterwards than to risk spoiling them. Remove the skin. Unjoint the neck close to the skull, enlarge the opening at the base of brain and remove all the brains. Take out the eyes and tongue and remove all flesh from the skull.

This completes the skinning, except when the animal has horns. In that case, proceed as above until you come to the neck. Skin as far forward as you can and unjoint the neck at skull. Then cut across from one horn to the other and loosen the skin around the horns. Then make another cut from between the horns backwards along the neck far enough to make the opening large enough to take the skull out; skin out the skull and the skinning is completed. The flesh must now be cleaned from the skull by boiling until tender, then scraping it, or simply by cutting and scraping it.

Take out the brains, eyes, etc., we are now ready to anoint the whole inside of skin, skull, etc., with the preserving powder. Put in the frame and stuff the body if ready; if not, the ears, nose, feet and lips must be well wet with the preserving fluid and laid away in a cool, dry place.

STUFFING AND MOUNTING ANIMALS.

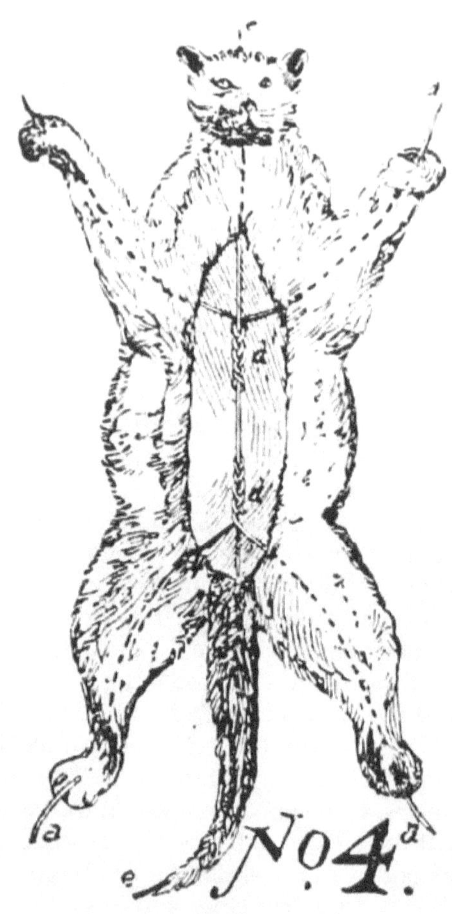

NO. 4.

After the skin is prepared, we will begin by preparing the skull. After cleaning as directed anoint it with the preserving powder, then put in suitable eyes, as directed in birds. Now if the legs are large enough to take in the leg wires, drill holes through the ends large enough to receive said wires. Next divide the skin of ears out to the points and place a piece of tin in them the shape of ear. Now turn the skin of head inside out and sew up the mouth, draw the under lip up under the upper lip and sew it there so as when turned back the stitches will not show. Next stretch the hide on the bench and measure it. In order to make the wire frame as shown and described in wire frames, Nos. 4 and 6, make the frame of suitable malleable wire, and place it in as shown, except the neck wire should be passed out at nostril. First put the skull in place and pass the neck wire through the nose.

Then pass the leg wires out through the leg bones, or on the inside of skin where the bone is not large enough to receive it. Now cut the tail wire long enough to reach to end of tail and put it in place.

Now, if the skin has not been previously prepared with some of the preserving preparations, powder the whole inside with the arsenic and alum. However, the head and legs have to be prepared before putting in the frame. We are now ready for the filling. Stuff cotton in around the leg wires to make the legs proper shape, being careful not to get it in too tight. Take some sand and mix some arsenic with it; fill the tail with this, placing some cotton at base of tail to hold the sand in. Now stuff the neck and body with any suitable material, shaping and sewing up as you proceed. Examine the head, and where it needs filling out pass the cotton in through nose and ears, and shape it as in life, arranging the ears, nose and mouth. Now pose the specimen as you wish it to remain, and place it in a cool, dry place to dry. The leg wires should be bent so as to allow the foot to rest on board, and when dry the wires should be cut off at bottom of feet, and the projecting neck wire at end of nose should also be cut off. Anoint the feet, nose and ears with the preserving fluid once a day for a few days. The wood and wire frame is used in heavy animals, the wood simply taking the place of the wire along the back. By examining the cuts it is easy to see how they are to be used.

Wishing to stand animals, such as squirrels, groundhogs, etc., on the hind legs, they have to remain on a block or perch; so procure a suitable block or perch, and varnish it before putting the specimen on. Animals and birds can be stuffed with hay, or anything that you can get into them in good shape. I often mow fine bluegrass and dry it carefully, and find it makes splendid stuffing, especially for the bodies of animals and large birds. Excelsior also makes a good filling.

ON MOUNTING DEER HEADS.

In mounting a deer's head to look life-like (having horns on), proceed to skin, anoint with the preserving powder, build up the nose with plaster of paris, put in the eyes, sew up the mouth and neck, for which see directions in another article. Now take a 2×4-inch piece of pine for a neck support. Fit the end of it in the brain pit with plaster of paris. When the plaster has hardened sew up the mouth as directed in cat. Draw the skin up over the skull and sew the skin together, drawing it close around the horns. The skin being opened on back of neck, sew it up to within six inches of back end. Now make a board to fit in back end of neck skin, and fasten it to the neck support in such a way as to hold the head on wall as desired. Tack the skin around the board, being careful not to draw the hair down. Stuff the neck and finish sewing up. Cut the skin off at back of board. Arrange the whole head as directed in the cat and set away to dry. See that the ears and eyes are all right, and use plenty of the preserving fluid. When the head is completed make a suitable shield to place on back end of the neck; this should be of a neat pattern and nicely polished. Fasten this on to neck board with screws, then on the wall, also with screws, which should be passed through the shield, under edge of hair, to conceal the heads of screws.

ARTICLE III.

A SIMPLE METHOD OF SKINNING, STUFFING AND PRESERVING FISH.

FISH DRESSING.

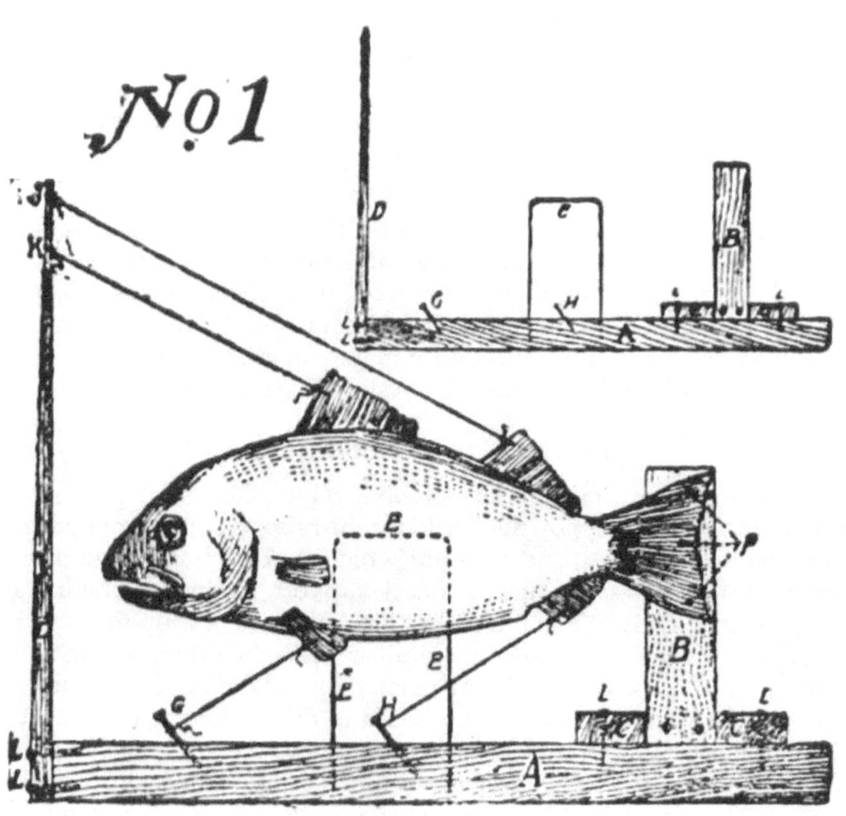

NO. 1.

Take a scale fish, say one that weighs two pounds, more or less, and if a fish on which the scales are tight, you can skin and dress it without losing any of the scales; but if a loose scaled fish, it will be better to protect them as follows: Take some tissue paper and press it gently on one side of the fish, then turn it over and treat the other side in the same manner. The natural glutinous matter which covers the scales will be sufficient to make it adhere firmly. Without this precaution the skin could not be removed from some fish without losing the scales, which would ruin the specimen. When these papers dry, lay the fish on its back, take a pair of scissors and open the skin down the center of the belly, beginning close up the gills and running clear back to the end of the body.

The skin should now be taken off with great care, using a smooth-edged knife for skinning, and a pair of scissors to clip the fins and other fine bones with. After the body has been skinned, take out the gills, tongue and eyes.

It is now ready for the preserving powder, which should be spread or dusted over the inside of the skin, being careful to reach all parts around the gills, in the mouth, eye sockets, etc. We now insert the frame E (see cut No. 1), which is simply a wire bent as shown in the cut, and sew up, beginning at the tail and ending at the gills, leaving the ends EE of frame projecting, as represented in cut No. 1. We now proceed to fill the body with plaster paris. Mix up a sufficient quantity, take a funnel, open the gills or mouth and insert the tube of the funnel therein, and pour in enough plaster paris to fill the body. Handle the body carefully, and shape it properly, while the plaster is setting, being careful to keep the frame EE straight with the body. When the plaster has set (or hardened), bore two holes in a rough board to set the frame rods EE in, which will hold the body up off the board. (See cut.) Now proceed to stuff the head with cotton, being careful not to press the gills out of shape, then draw the mouth together with a thread and tie it. Fill the eye socket with plaster paris or putty, and set the eye in the proper position.

We are now ready for the fins and tail; if they have become dry, wet them with water and spread them out; then take a needle and thread and draw them up, as shown in cut No. 1.

To make the frame, take a rough board, say one inch longer than the fish you intend to dress. For a fish a foot long, the board should be one inch thick and about three inches wide, and for larger fish in the same proportion, or large enough to support the fish. To one end of the board nail a piece of lath a foot long (see D in cut); on the other end nail a block (C in cut), to which nail an upright piece (B in cut). Tack the tail to the upright (B) with tacks (P). Raise the upper fins by drawing a thread through them and tying to upright piece, as shown in cut. Drive two nails in board (A), as shown (GH), to which fasten threads attached to lower fins as shown in cut. Set away to dry and when thoroughly dry wash the whole body, take off the tissue paper. When the fish gets dry, dampen all the outer surface with corrosive sublimate and set away. Take off the threads and place it on a nice, smooth board and give a light coat of varnish. The specimen is then ready for the cabinet.

Should you prefer to stuff the body with cotton instead of using plaster paris, proceed as follows: After skinning as directed, make a frame, EE, and take a straight piece of wire long enough to reach from the point of the nose to the other end of the body, fasten this to the frame, EE, at the upper end, by wrapping them together with a small wire or cord, leaving the frame EE as near the center as possible. Take cotton or tow and wrap the wire that is to run lengthwise of the body, say one-fourth as big as the body, then insert it in the skin and proceed to stuff with cotton, being very careful to get the natural shape. When this is completed, proceed to sew up and mount as directed when dressed with plaster paris. The Gar fish makes a fine specimen, by simply filling the skin with dry sand while drying, and let it out when dry.

ARTICLE IV.

HUNTING SPECIMENS AND CARE OF SAME.

In hunting specimens, birds or animals, it is best to take a double-barreled shot-gun, have your cartridges loaded some with fine shot, and some with coarse; keep one barrel loaded with one number and the other barrel with the other number, then you are ready for either small or large game. Have some cotton and old newspaper with you, and as soon as you kill a specimen, stop up the shot holes with the cotton, and it is sometimes best to stuff some in the nostrils and mouth. If any blood has run out on the hair, or feathers, as the case may be, wipe it off carefully, then wrap the body in paper and pack it in the game sack, using all care not to bend, break or soil the feathers, for if once soiled it is hard to make the work look well again.

On removing the birds from the game sack, if there is any blood on the feathers, wash it out with clear water, and wipe until nearly dry; then sprinkle with plaster of paris and shake until dry. In this way the feathers will look smooth and natural; but if, after washing, you let them dry without stirring, they will present a bad appearance. Where the blood has remained on the feathers they must be well washed and treated as directed in a test case as given at close of Article I.

ARTICLE V.
WOOD AND WIRE FRAMES.

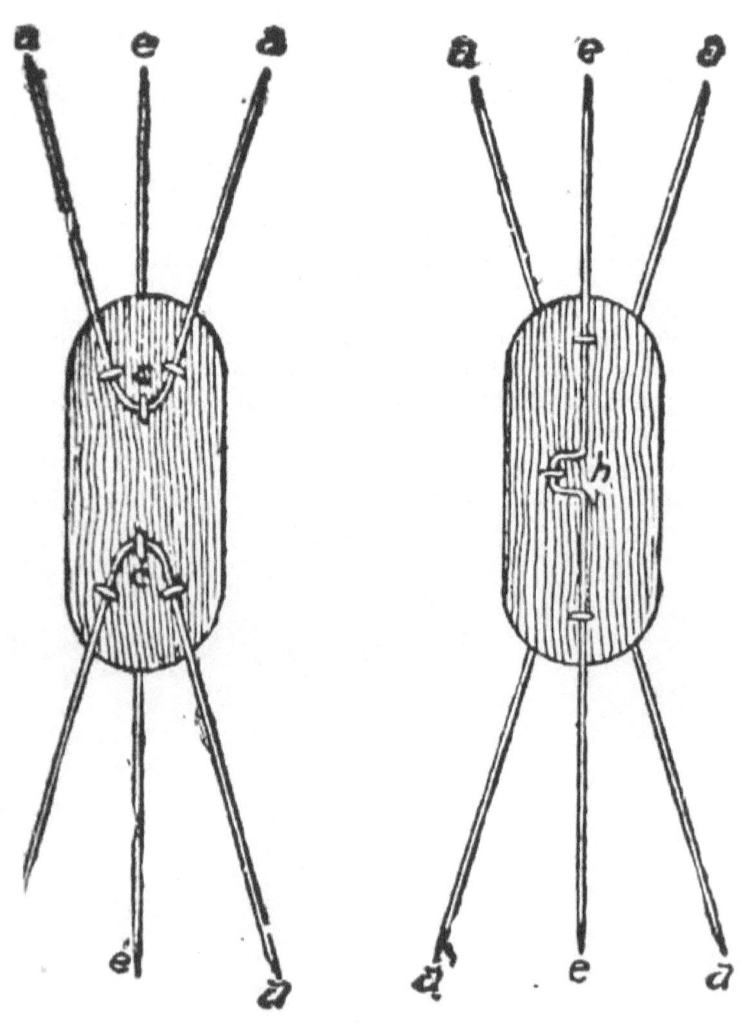

FIG. 7.   FIG. 8.

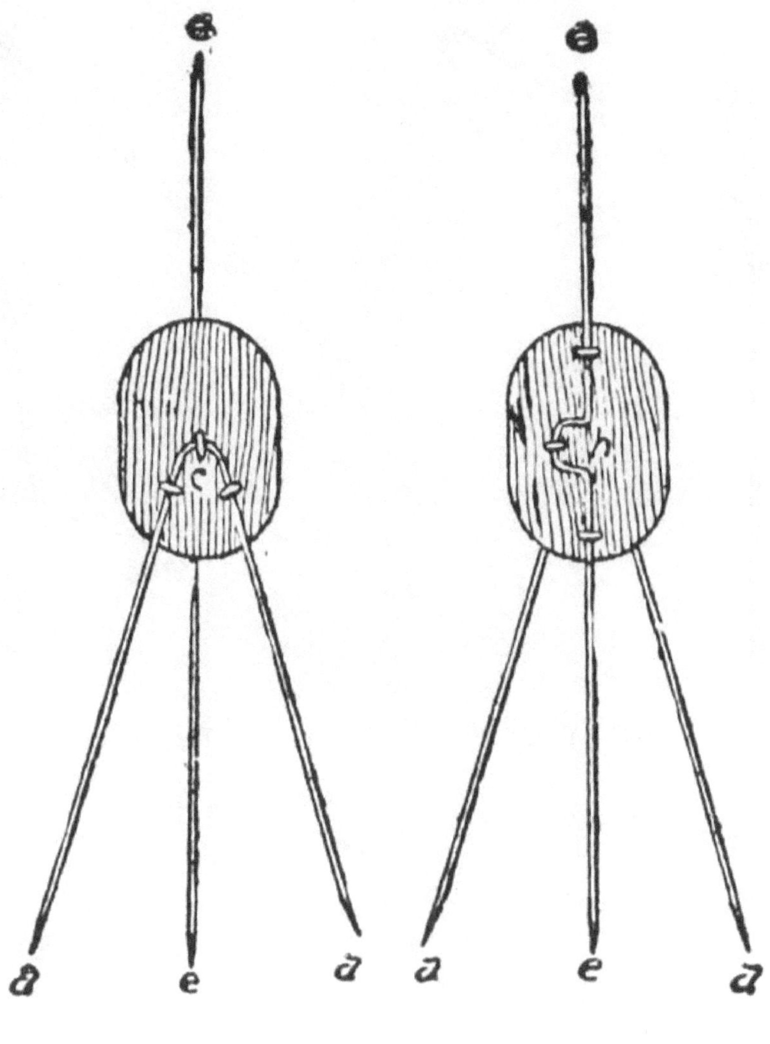

FIG. 9.   FIG. 10.

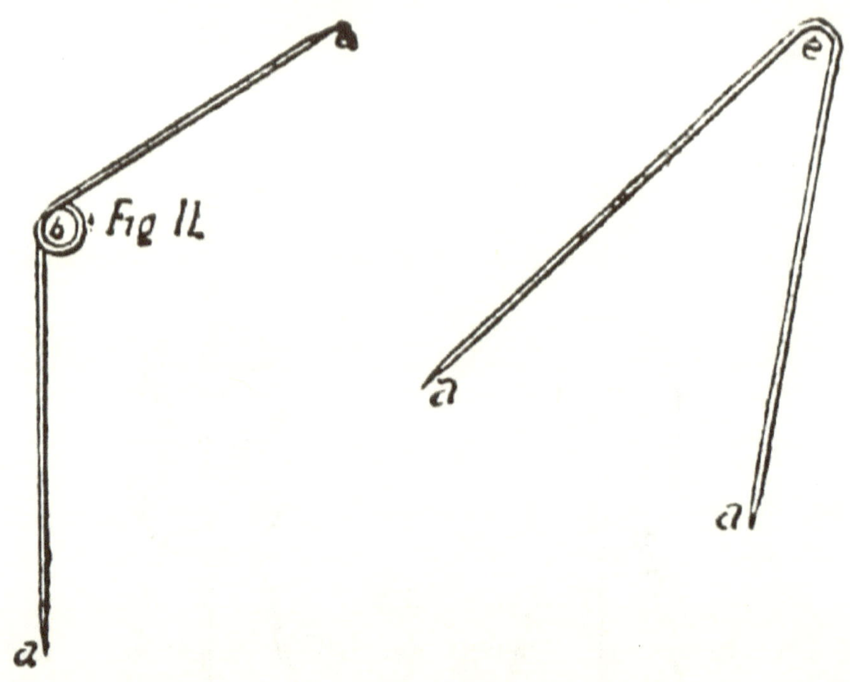

FIG. 11.   FIG. 12.

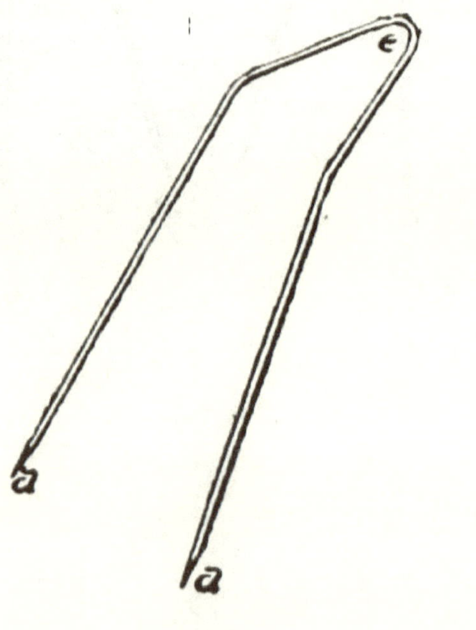

FIG. 13.

The above cuts represent my wood and wire frames for animals and birds.

Fig. 7 is Fig. 8 reversed; Fig. 8, showing the way the neck and tail supports EE are fastened to the body board with blind staples. Fig. 7 shows the way the leg wires AA are fastened in the same manner. In making the frame, measure the animal from shoulder to hip, and cut the board a little longer than the measure taken; round the corners as shown in cuts, and round the upper edges to keep them from cutting the skin should it come in contact with it. Now take a wire or rod long enough to reach from the end of the tail to the end of the nose, bend the wire as shown in Fig. 8, and fasten it to the body board with blind staples, unless the specimen be a large one, then it will be necessary to put them on with strong wire staples. Next bend the wires for legs as shown in Fig. 12, which should not be fastened to the board until after being placed in the legs of the animal being dressed. They are then bent up as shown in Fig. 13 and fastened to the body board with staples to suit.

The frame for birds is made in the same way, excepting the forelegs, which is used on bird frames only when wanting to spread the wings. Fig. 11 is a single wire frame, to be bent in the manner shown, and to be used in dressing small birds. (See small bird page 5).

ARTICLE VI.
FOR BIRDS AND ANIMALS.

WIRE FRAMES.

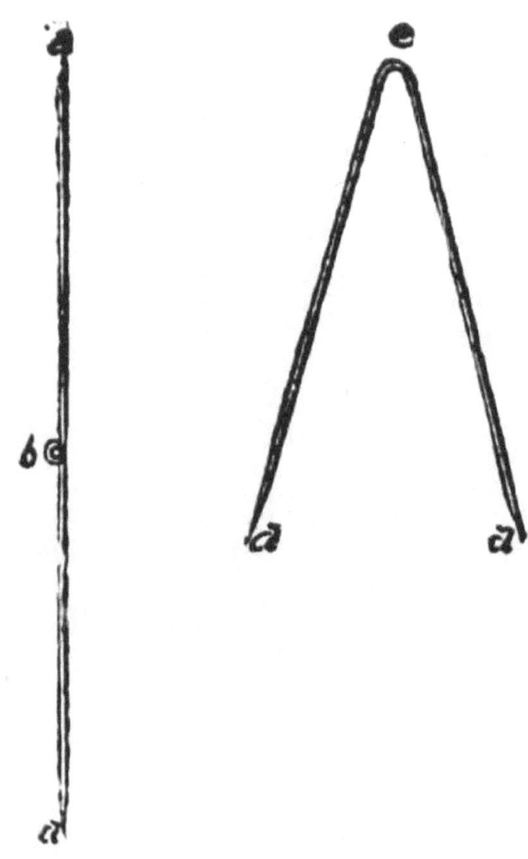

FIG. 1.   FIG. 2.

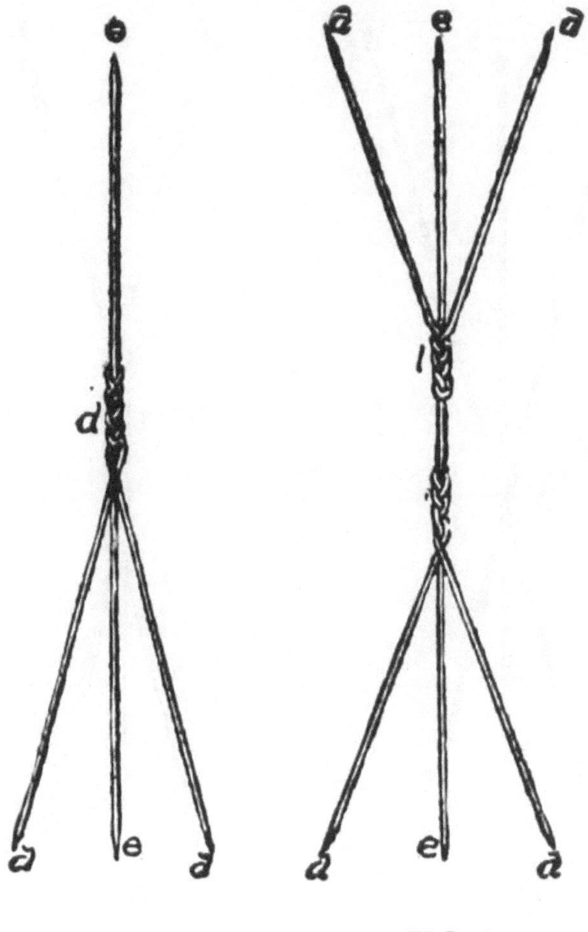

FIG. 3.   FIG. 4.

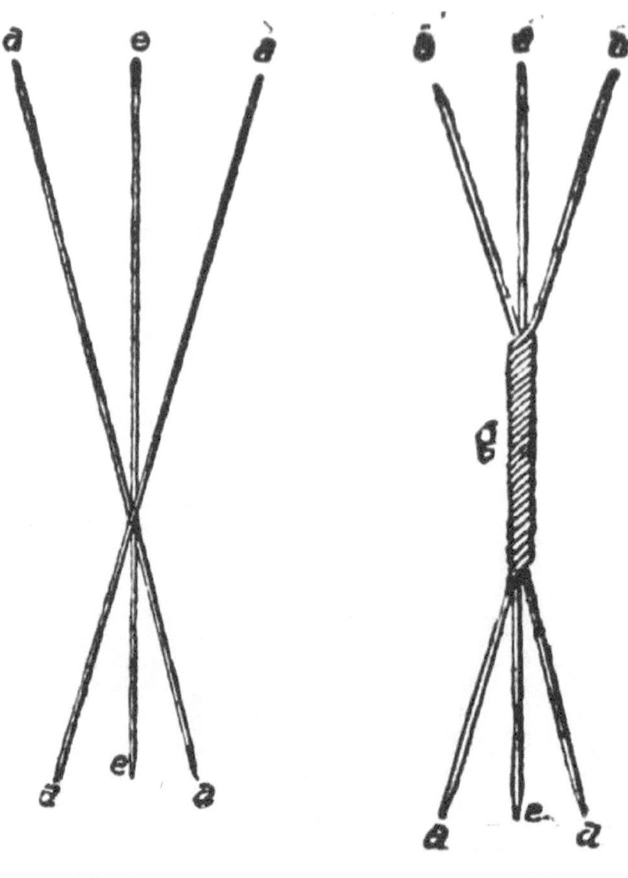

FIG. 5.   FIG. 6.

The above frames are made as follows: Take a straight wire and turn a loop in the center as shown at B, Fig. 1. Take another wire about the same length and bend it as shown in Fig. 2. Now run Fig. 2 wire through the loop B in Fig. 1 down to bend E, fasten them in a vice at the loop B in Fig. 1, and plait the three together as shown in Fig. 3. To form the Fig. 4 frame, make another loop two or three inches or further up the neck wire E, bend another wire as Fig. 2, put this through the loop, and plait together as in Fig. 3. This forms the frame for animals, as shown in Fig. 4.

For Figs. 5 and 6, lay three wires together as shown in Fig. 5, put them in a vise and twist or plait together as shown in Fig. 6. This forms three wire frames—Figs. 3, 4 and 6.

The advantage Fig. 4 frame has over Fig. 6, is that it is easier bent in the center, there only being one wire.

ARTICLE VII.

MOLES.

It is not necessary to give more than one way to dress a mole. An expert can skin, stuff and sew up a mole in ten minutes in the following manner:

After killing a mole, let it cool; then make an incision along the abdomen, commencing at the tail and running forward two inches, or about to the center of the body. Skin out the hind legs, bone and flesh, down to the foot and unjoint. Press the skin back to the tail and unjoint it close to the body. Pull the skin off the body, taking care of front feet, not to pull them loose from the skin; unjoint them, or break the arm bone close to the foot; skin on down to mouth and cut the body loose, leaving only the jaw bones in the skin. Spread on the whole skin all the preserving powder that will stick to it, and turn the skin back inside out. Turn the hind legs and proceed to fill the body with cotton, using a wire to stuff the cotton in with, as in all other animals, being careful not to stretch the skin, and also not to get the cotton knotted up. After the body is filled sew it up and place on a board to dry. Place the feet in position, and wet them and the tail and nose with the solution of corrosive sublimate.

Other small animals, such as rats, mice, ground squirrels, gophers, etc., can be dressed in the same manner. After stuffing smooth the hair and shape the body before laying away.

ARTICLE VIII.

MODE OF DRESSING AND PRESERVING TURTLES.

A good sized turtle may be thoroughly preserved without the tedious process of unjointing, skinning, digging out the flesh, sewing up, etc., by observing the following directions:

Take a snapping, or any other hard shell turtle, say one that measures ten inches across the back, or smaller, down to the smallest.

The first thing to be done is to kill it, which is the hardest part of the operation, as it clings to life tenaciously, and large ones are very strong and hard to manage. One way is to get it to open its mouth, and catch it by the under jaw with a pair of strong pliers, hold its mouth open and make it swallow a quantity of the solution of corrosive sublimate; this being a deadly poison, it will soon kill it.

Another way is to hold its head as far out as you can pull it, and stick it as close to the under shell as you can, running the blade well back into the body. It takes some time for it to die after being stuck.

And still another way is to open its mouth and stick a knife blade up through the roof of the mouth into the brain.

I prefer the poisoning, as it seems to die easier and quicker, and it does not disfigure it. As soon as it dies, lay it on its back, and make an opening in the skin, under one of the back legs, large enough to take out all the entrails; now take a pair of pliers or a hook (made for the purpose); pull out all the entrails, liver, heart, etc. If the specimen is a large one, pour in some water and rinse out thoroughly, then put in a lot of the preserving powder, being careful to get it well distributed through the body. Then take some cotton and dust it well with the preserving powder, and fill up the inside, pressing the cotton in tight, filling the body full and sew up. Now open the mouth and pour some of the solution of corrosive sublimate down its throat, and press some cotton sprinkled with the powder down its throat, being careful not to stretch the neck out of proportion. Remove the eyes and insert the artificial eyes in their stead; fill the mouth with cotton and close it. Cut a hole in the bottom of each foot, and probe the legs with a knife or the blunt end of a wire; pour in some of the solution of corrosive sublimate, or work in enough of the powder to preserve the flesh; if the tail is a large one, it should be treated in the same way, and all sewed up. Now set it on a board and tack the feet to the board in as near the way it would hold them while walking as you can. Pull the neck out a very little, and put a little block or a small roll of cotton under it to hold it up in position while drying. Before putting it on the drying board, take a small paint brush, pour out some of the solution of corrosive sublimate in a dish and give the whole outer surface a thorough wetting. This should be repeated once a day for several days. This completes the dressing. After the specimen is dry, take it off the board and set in the cabinet.

Another way: After killing, as directed, lay the specimen on its back and un-joint the breast shell from back shell with a chisel or heavy knife blade; open the skin from the front part of the foreleg back and around to front of the other fore-leg, cutting under the legs. Skin out all the flesh and bones of the legs and unjoint them at the foot, skin out the tail, then unjoint the neck bone from backbone, and skin it out and unjoint at the back of the skull. Take the flesh from back and breast shells, then anoint all parts with the preserving powder. Take a frame as Fig. 8, in article V, of the wire and wood frame, and put in legs, neck and tail, as directed in animals, letting the neck wire pass out through the mouth or nostril. Stuff the legs, neck and tail with cotton or tow, and sew up the sides, leaving an opening to stuff the body; fill the body, then finish sewing up. In stuffing the legs, neck and tail be careful to keep the wires in the center of each. Open the mouth and take out the eyes from inside and put in the glass eyes; put some cotton back of them to hold them in place while drying. Close the mouth, and wet the whole body with the solution of corrosive sublimate. This should be done once a day for several days. Place the specimen on a board to dry. For position, copy after nature as near as possible. After the skin is dry a coat of varnish adds much to its appearance.

ARTICLE IX.

KILLING AND DRESSING SNAKES.

To kill a snake without bruising or breaking the skin is a difficult undertaking, for, as a general thing, we are not looking for snakes, therefore we are not prepared to capture one.

When we do run across them, we generally kill them with the first thing we get hold of. If the snake is not too large, strike it across the back with a small stick. This disables him, and you now have time to prepare to finish him. Watch your chance and when the opportunity presents itself, tap it on the head a few times, and you can soon kill it without bruising. When it is dead, open its mouth and cut the tongue and fangs out, then unjoint the backbone from the skull and skin back on the body an inch or two, turning the skin inside out. Tie a cord to the skinned part of the body and pull on this with your right hand, while with your left hand you pull the skin off from the body, which is easily done if the snake is not too large in the center of the body; in that case it must be opened in the largest part of the body. Make the opening on under side of body, lengthwise of the snake, and long enough to allow you to remove the skin all around the body, then cut the body in two. Skin out both ends, unjointing the body close to the skull, turn the skin right side out and sew up the opening, taking short stitches. Now mix some plaster paris and water, leaving it thin enough to run; place a funnel in the mouth and pour the plaster in until the body is full. Lay the snake on a level board and coil it before the plaster gets dry, placing the head in the position in which you want it to remain. Take out the eyes and put in the glass ones; then wet the whole skin with the solution of corrosive sublimate. This should be repeated once a day for three or four days.

Another method is to skin and sew up as directed, then take a long wire and stuff the body with cotton or tow, being careful to get the stuffing in smooth, so that no lumps will show on the outside.

Another way is to fill the body with sand, and when the skin has thoroughly dried, make a hole in the under part of the body and let the sand run out.

In either of the above modes the snake should be coiled as soon as stuffed in the shape you want it to remain, and plenty of the solution of corrosive sublimate used over the skin.

Eels, frogs and toads are skinned and stuffed in the same manner as the snake, no frames being used.

ARTICLE X.

DRESSING ALLIGATORS.

Take an alligator, say five feet long, make an incision the full length of the body, on the under side, and skin out all the flesh and bones, as recommended in skinning animals. Scrape off all the fat or flesh that may be left on the inside of the skin, and give it a thorough coat of the preserving powder. Now proceed to stuff it without using any frame. First stuff the legs with cotton or tow, pressing it in very tight—their skin being very tough, there is no danger of stretching it. Sew the neck up, down to the front legs, and stuff tight; now sew up the tail to hind legs and stuff it; then take a piece of pine board, say an inch thick, two inches wide, and twelve inches long, place this inside of the skin, draw the edges together over it, and tack both edges close together on the board, fill the body up to the end of this board, and put in another board in the same way, and again stuff; when the full length of the opening is closed up in this way, before putting in the last section, have the body thoroughly stuffed and put the section in afterward.

The advantage in putting in the board is this: It is almost impossible to sew the skin, and as it is to lay flat, it is much easier to use the boards; they could also be used in the neck and tail, if you wish to lay them straight with the body. Lay the specimen on a flat board, and place his feet and tail as you wish to have them. The eyes should now be removed, some of the solution of corrosive sublimate put in the sockets, then put the glass eyes in with putty or plaster paris. Close the mouth and set a block under the jaw to hold it up while drying. Give the body a thorough wetting with spirits of turpentine, repeating it once a day for three or four days. After the skin becomes well dried give it one or two coats of varnish and you have a fine specimen.

ARTICLE XI.

DESCRIPTION OF FOOT-STOOL.

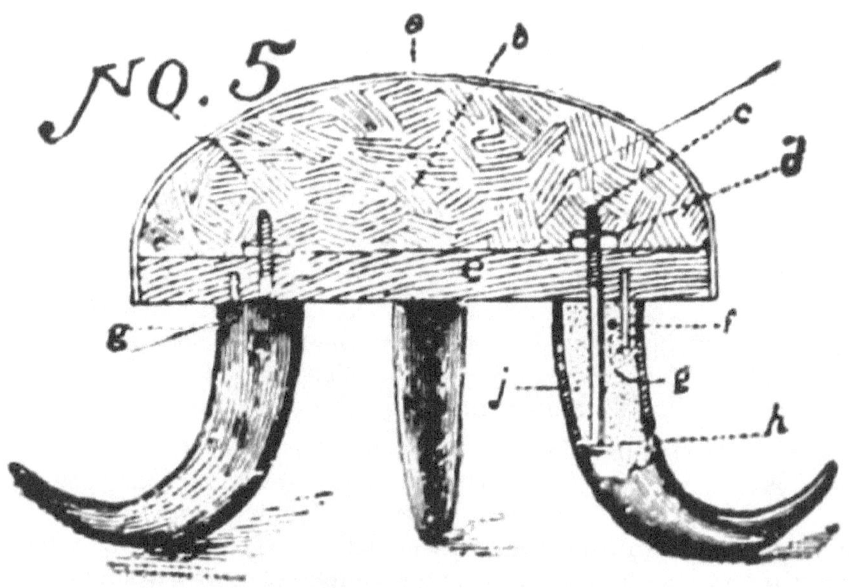

NO. 5.

Letter (A) represents the covering, (B) the moss or hair filling, (C) the bolt which fastens the horn to the stool brace, (D) the tap of said bolt, (E) the wooden base of stool, (F) a small rod or nail set in the plaster paris, (G) a rivet through the horn, (H) head to bolt, (J) plaster paris in the horn.

DIRECTIONS FOR MAKING THE STOOL

Take three cow horns of good shape and size; rasp them and sand-paper down, as hereafter directed in Article XII; then, after deciding what height you want the legs, make a frame to saw them off on, as illustrated and directed in Article XII, cut No. 6. After sawing off, drill a hole through in the horn about three-quarters of an inch from the large end. This hole should be large enough to take in a ten-penny nail. Put the nail through and file it off even with the horn on both sides, and rivet it with a small riveting hammer. Now level up the ends of the horns where they were sawed off, by taking a coarse piece of sand-paper, and laying it on a level board, rub the ends of the horn round and round on it until it is perfectly level. Now take the bolt C, of size and length to suit the horn, put it down in the horn to see that it will suit; sometimes it has to be bent in the shape of the horn; after fitting it to the horn, leaving about 1¼ inches projecting out to go through the wooden base E and leaving room for the tap D. Now mix some plaster paris and water until about as thick as it will run; pour this in the horn until nearly full, put in the bolt, as shown, and while the plaster paris is yet soft, take an eight-penny wire nail and push it down head first in the plaster paris, leaving about half an inch sticking out; this nail and the bolt should both come straight up from the level of the end of the horn. When the plaster is partially dry, level up around the bolt and nail at the end. The horn should now be polished as directed in Article XII. The next is the wooden base E; have a board turned or cut round, say twelve inches across and one inch thick. This should be made of walnut, or some other hard wood, but when the hard wood can not be procured pine will do. If pine is used it should be painted, or stained and varnished on the under side, and if walnut it should be sand-papered, filled and varnished. Now strike a circle within about two inches of the outer edge and bore three holes at equal distances apart and set in the bolts; when you get them in the position you want them, press down on them to get the imprint of the nails F, then remove the horns and bore a hole the size of the nails F. Now put the bolts in again and screw the top on tight. The nail F is to keep the horns from turning around and getting out of place. See that the butt of horn fits tight to the wooden base E.

UPHOLSTERING FOOT-STOOL.

Go to a saddler and get either curled hair, deer hair, or moss sufficient for a heavy pad; lay it on the floor and beat all the dirt out of it, then pick it all apart to get all the matted lumps out, lay it on the stool top and pile it up about a foot high while loose, take a piece of muslin or drilling, lay it on top of the moss or hair and draw all the four corners down tight, tacking them about the center of the edge of the wooden base E, then draw it all down and tack in the same manner, drawing very tight, being careful to get it round and even and leaving no wrinkle. In case the moss is not evenly distributed, take a long awl and run it through the covering and carefully distribute it.

Next in order is the outside covering, which should be a fine piece of silk plush or velvet. Get a piece large enough to work some fine flower or motto in the center, then put it on in the same manner as the first, only tacking it down near the lower edge of the wooden base E; by being careful you can get it on without wrinkling it. The tacks should be small and very close together. Put a piece of braid around the edge and tack it on with upholstering tacks and the stool is completed.

SAWING OFF HORNS.

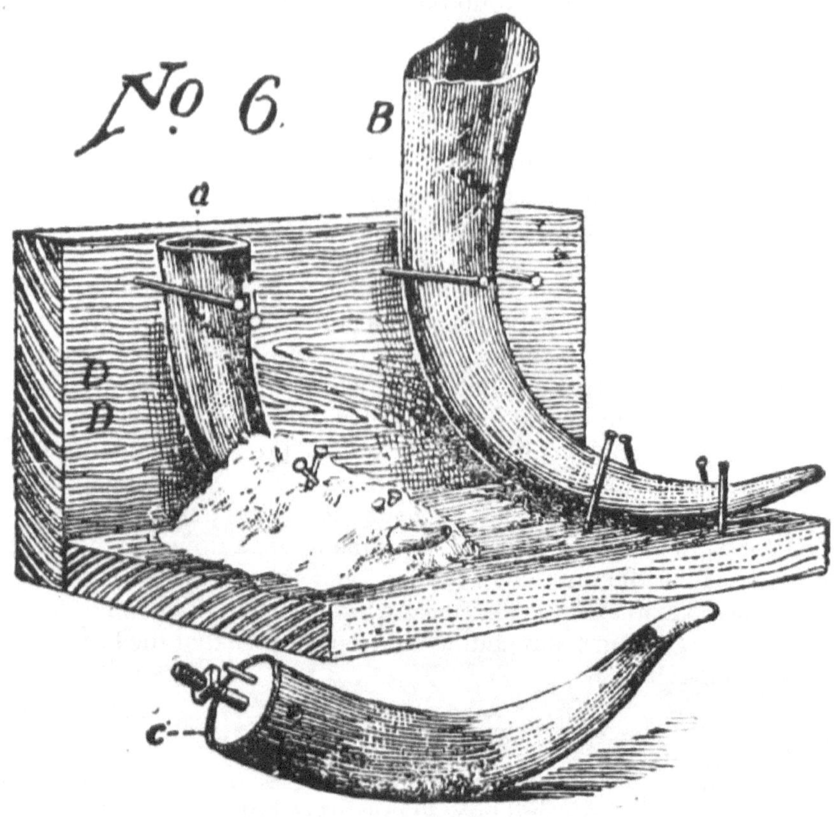

NO. 6.

Cut No. 6 is to illustrate a frame on which to cut horns to be used for footstool legs. Without a frame of this kind it is almost impossible to cut them to set level with stool and floor. In fact, it is more difficult than it would be for a carpenter to cut molding to fit corners, etc., without a mitre box. It is very simple and is made as follows: For a frame large enough to cut four horns, take a fencing board and cut off a piece from it two feet long. The board is supposed to be about six inches wide and one inch thick. Saw off another piece the same length and nail it to the edge, as shown in cut. This we will call the back of frame and the other the bottom. The back then is five inches high, measuring from bottom board up. This completes the frame as shown. The horns are now placed on the frame and nails driven around them into the frame to hold them firmly in place. Place the horns close together in setting on the frame and pour plaster paris over the points of all, as shown in cut horn A. After the plaster paris has thoroughly hardened, saw off another piece of the same board and nail it to front edge of bottom board, and proceed to saw off the horns level with the upper edge of back and front boards, as in horn A. Should you want the legs longer or shorter, take wider or narrower boards for back and front of frame. After taking the horns from the frame, proceed to put in the wire through the butt of horn and rivet it; then put in the bolt and nail F (as directed in a former article) and fill the horn with plaster paris, as shown in the above cut (C). The horn is now ready, after polishing, to be placed on the stool. It can also be placed on a board and used for a hat hook, or for holding back window curtains, etc.

ARTICLE XII.

DIRECTIONS FOR POLISHING HORNS, HOOFS, ETC.

Go to a hardware store and buy two good wood rasps, flat on one side and round on the other; then get three grades of emery or sand-paper, say No. 1, No. 0, No. 00; then go to a drug store and buy ten cents' worth of sweet oil and ten cents' worth of rotten stone. Get an old felt hat and an old case knife and you are ready for business.

Take of either horns or hoofs, several at a time, and put them in an old pot or kettle of boiling water, let it boil until they become soft, which usually takes about half an hour; take out one at a time and rasp it until it begins to cool and harden, put it back and take out another, and keep this up until all are rasped down to the shape you want them. Now take one at a time in the same manner and scrape them until they are scraped smooth of all rasp marks, using the case knife for the scraper, the edge of which has been ground square off like that of a shear blade. Now take the No. 1 sand-paper and sand-paper down, then use the No. 0, and finally finish up with the No. 00, rubbing-lengthwise of the horn. We are now ready for the finishing touch. Take a piece of felt sufficiently large to rub with, pour some of the sweet oil on it, then dust with the rotten stone, and give the horn a thorough rubbing, putting more of the rotten stone on occasionally, and keep up the rubbing until you get a high polish. Such a polish will remain on for years.

Another way to polish horns: Fasten them firmly on a bench or to the wall. Take a strip of ticking or felt about three inches wide and three feet long. Wet it in water and sprinkle powdered emery over both sides. Take hold of both ends, place it around the horn and draw backwards and forwards, around and around, until the horn presents a smooth surface; then take another strip of the same description and use in the same manner, except using rotten stone instead of the emery. There are different grades of the emery, and only the finest is to be used in polishing; a coarser grade can be used in the same manner in the place of sand-paper, and is very good.

COLORING HORNS.

If you wish to give a horn a flesh color, secure a very light colored horn, rasp and scrape it down very thin, and polish so that the horn is transparent. Get some red paint (with plenty of dryer in it) give the inside of the horn a coat of the paint, and set it away to dry. Before applying the paint, wash the horn out well with soap and warm water and dry well. The above gives a beautiful flesh color, which will last for years, especially when filled with plaster paris.

TO REMOTE PLASTER PARIS FROM A HORN AND TO MAKE THE HORN ROUND.

After the plaster paris has become thoroughly dry it is very hard; therefore to cut it out with a chisel is very tedious; but it often has to be done. If the inside of the horn is smooth and a little tapering, it is very easy to remove by boiling it until the horn expands or becomes soft; then tap the horn with a smooth stick or hammer, and the plaster will drop out in a whole piece. If you have put a rivet through the horn before putting in the plaster paris it is more difficult to remove, as the rivet holds the plaster firmly in the horn. In such a case take a chisel and dig out the plaster paris down to the rivet, then either cut the rivet with the chisel or file it in two and take it out; then proceed as above directed.

To make the butt of a horn round, put the horn in boiling water; let it remain about ten minutes; have a round wooden plug ready, which should be wedge shaped; drive this in the horn while hot, and let it cool; afterwards remove the plug, and the horn will remain round. This makes quite a difference in the appearance of the horn where you wish to use it for a hat rack, footstool, etc.

TO DRESS DEER HORNS.

Make an extra strong lye, using the granulated lye and hot water, say one tablespoonful of lye to half pint of hot water. Give the whole horn a thorough bath in the mixture, and let it set about one hour; then take a stiff brush and give the horns a thorough washing; this cuts off all the dirt that has accumulated in the rough knotty parts. The horns should now be put in a barrel of water and let soak a day or two, to take off all the lye, then wash well using a stiff brush. Now take an old hand-saw and cut down through the skull, commencing one inch back of the horns and coming out at the eye sockets, being careful to saw straight through, so that when placed on a board they will set level. Take a two-inch pine block as wide as the skull and three inches longer, round off the top and bottom ends to suit the shape ox the skull, place the horns on the block, holding them there with the hands. Put it up against the wall and if the horns pitch too far forward take down and cut off from the front of the board, sloping it backwards until the horns hang at a proper pitch. Now drill about four holes through the skull, and screw the skull fast to the board. Take a dozen or more shingle nails and drive in the front of board an inch or more apart, leaving the heads sticking out. These are to hold the plaster paris, which is put on to level up before putting on the covering. After the nails are all in, mix some plaster paris and pour over the skull and around the edges to round it up in shape for the covering. After the plaster paris is partially hardened, smooth it by scraping down the high places, then let it dry. Take a piece of strong unbleached muslin large enough to stretch over the skull and board, tack the muslin on the back part of the skull board, then draw it down between the horns and tack it to the under side of the board at the lower end, drawing very tight; now commence at the edge of the muslin and cut straight in to the inner side of horns, round out a little for the horn, and proceed to draw it all down tight, and tack it on the under side of the block, being careful not to make any wrinkles in the covering; trim it off to the tacks on the under side. (Some prefer to have the muslin wet while putting it on.) We are now ready for the outer covering, which should be plush or velvet, and is put on in the same way as the first covering. It requires some skill to make a smooth job of the first; yet by going slow, any one should do a fair job. Next make a board to screw on the back of this, which should be made in diamond shape or rounding on both ends, and about six inches longer and one inch wider than the skull board; this should be made of walnut and nicely varnished; after it has thoroughly dried fasten it tight to skull board with screws, counter-sinking their heads. Bore one hole at each end of the base board and counter-sink them; then screw it to the wall. Varnish the rough part of the horns and let them dry; then take some chenille or tinsel cord to trim around the butt of the horns and around the back part or edge of the skull board. Tack it on with upholstering tacks. Cow horns and sheep horns are mounted in the same manner. The sheep horns should be cleaned with the lye, as directed for the deer horns, but no scraping or polishing is necessary; they look better rough. The cow horns should be polished, as directed on page 102, before they are put on the board. No polished horn should be varnished.

POLISHING HOOFS OF VARIOUS KINDS.

The cow hoof takes a high polish, and a number of useful articles can be made of them—which are not only useful, but quite a novelty. Take a pair of cow hoofs, which are neither bruised nor scaled, rasp them down in good shape, then scrape, sand-paper and polish the same as directed in polishing horns. The hoofs should be good matches, and both from the same foot. Bore or drill two holes in one of them on the inner side, about one-half inch apart, and place the two hoofs side by side on a level board, seeing that the toes are even with each other. Take a piece of wire or an awl and put through the holes in the hoof and mark the place to make the holes on the other, so that when they are wired together they will set even with each other. Drill small holes around the tops, say a quarter of an inch apart; take a piece of malleable wire of sufficient size and draw it through the holes in the sides of the hoofs, drawing them firmly together, and twist the ends of the wire together; this will hold them in place. Bind the upper edge of hoofs with piece of cloth, sewing through the holes drilled for that purpose. Stuff each hoof with tow or moss, and stuff with any goods to suit your fancy, silk plush, velvet, etc. Put the filling in tight and let it come above the level of hoof to make the cover rounding, then drawn down tight. Draw the edges of the cover down and sew them with a strong thread to the binding. Take a piece of fancy cord or ribbon and tack it around the edge of the cover and you have a fancy pin cushion. To make a match safe or tooth-pick case, polish the hoofs, wire together, and drill the holes around the edge as before; then take some glue and spread it well over the inside of each hoof; cut some red flannel to fit the inside and press it in with your fingers until it is all smooth; cut it off at the edge of top of hoof, and bind the edges with ribbon. This makes a very pretty match safe or tooth-pick case. All hoofs are polished in the same manner. Should the hoofs not set level after they are put together, take a coarse piece of sand-paper and tack or hold it on a level board, set the bottom of the hoofs on it and rub round and round until they do set level. It would be best to attend to this as soon as they are fastened together.

ARTICLE XIII.

ACORNS, TIPS FOR HORNS, ETC.

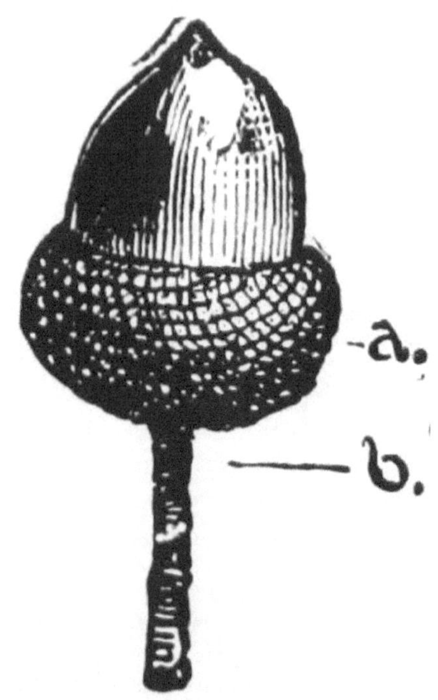

A nickel-plated acorn, as shown in the accompanying cut, makes a very fine finish on the point of a polished horn. B is a wire stem attached to the acorn, and is intended for fastening it to the horn. They are made as follows: If made of metal, they are cast on the stem and then turned. If brass, they are polished; but if iron, turned and then bronzed, painted or nickel-plated. If made of horn, first put in the stem and then turn and polish them. To put them on the horn, saw off the point of the horn, drill a hole in the end of same, the size of the stem, put a little glue in the hole and drive the stem in tight. The horn should be polished before putting the knob on. Round knobs are also very nice, put on in the same way. This way of putting them on is much better and easier than the old way of screwing them on.

ARTICLE XIV.

RECEIPTS FOR VARIOUS PREPARATIONS USED IN THE PRESERVATION AND SETTING UP OF ANIMALS, FOWLS, ETC.

SOLUTION OF CORROSIVE SUBLIMATE.

Corrosive sublimate	1 teaspoonful
Alcohol	½ pint

Mix and let stand twelve hours, and it is ready for use.

ARSENICAL SOAP.

Arsenic in powder	3 pounds
Salts of tartar	1 pound
Camphor gum	6 ounces
Cocoanut oil soap	1½ pounds
Powdered lime	5 ounces

Cut the soap up fine and put it in a kettle containing one gallon of soft water and boil until the soap is well dissolved, then add the lime, salts of tartar and arsenic, stirring the mixture thoroughly; put the gum camphor in a half pint of alcohol and place in a mortar and crush it; take the kettle off the fire and add the camphor; mix well, and when partly cooled put it in fruit jars and seal up. It is now ready for use, and great care should be taken in using it, as it is a deadly poison.

PRESERVING POWDER.

Arsenic in powder	2 pounds
Alum in powder	1 pound

Mix thoroughly and keep in a dry place. This is also very poisonous.

CARBOLIC ACID.

The following solution of carbolic acid and water I have found very good in preserving skins, bugs, etc.:

Carbolic acid	½ ounce

Soft water   2 quarts

Turpentine is also a good preservative, and is sometimes used in place of corrosive sublimate.

THINGS WORTH REMEMBERING.

Never attempt to dress a bird when its feathers are broken or badly blood stained, unless it is a rare specimen.

Never dress a bird after the feathers begin to slip.

Never dress an animal after the hair begins to slip.

Never let a hide get dry before applying the preserving powder.

Never stretch a hide out of shape while taking it off.

Never forget to stop all wounds and the mouth and nostrils with cotton as soon as you kill the specimen.

Never let a specimen get dry after stuffing it before shaping and setting it up.

Never inhale the preserving powder, or get it in cuts or sores, for it is a deadly poison.

Never be afraid of putting too much of the preserving powder on a hide.

Never be afraid of charging a good price for a well dressed specimen.

Always use the best material, such as eyes, tow, cotton, wire, etc.

Always try to improve on each specimen you dress that you may excel others in the art.

TANNING HIDES WITH HAIR ON.

Very little has ever been written in regard to tanning skins with the hair on. Indeed we may say there is scarcely any literature on the subject, and this article must necessarily be very brief. Tanning with the hair on is always somewhat difficult, but of course, some hides are more difficult to tan than others. As an average, I will here give directions for tanning a dog hide. The same process obtains in the tanning of other hides. The only difference being in the length of time required which must be regulated to suit each case. We will start with the skin on the dog's back, and hence the first operation will be the skinning process. To begin, make an incision from middle of under lip back along the median line to the tip of the tail. Then make an incision from the middle of each foot along the inner side of the leg to the median line, and proceed to skin, being careful not to cut holes in the hide, and at the same time to leave no flesh sticking to the skin.

When skinning the head cut the gristle of the ear back close to the skull and separate the skin as near out to the point of the ear as possible. Unless this skin is separated, it is difficult to tan the ears without losing the hair, especially in warm weather. Be very careful not to cut the eyelids; skin on over the nose and cut loose from the body.

Then go over the pelt and remove every particle of flesh which may have been left on the skin, and be especially careful around the nose and mouth. All the bones of the feet should be removed except the bones of the toes. Next bathe the nose, ears, feet and tail with the preserving fluids used in taxidermy. If you have none of the fluid, use plenty of arsenic and alum, and then stretch the hide on a floor with the hair side down and tack all the parts so that they will be stretched perfectly smooth, after which salt all over thoroughly. Let it remain until perfectly dry, and then take a piece of coarse sand-paper and rub it down smooth and clear of any remaining vestige of fleshy particles.

Or instead of the preceding modus operandi, remove the pelt from the floor when about dry, and lay it on a fleshing beam and with a fleshing knife scrape off all remaining flesh and grease. Should there be grease that cannot be removed with fleshing knife, tack hide back on the floor and spread about three gallons of sawdust over it and rub thoroughly. Clean the skin and cover again with salt, and when dry sand-paper as directed above. Next place in warm water to soften it, and when soft wash it thoroughly with soap. Then rinse and wring it, and put it into the following liquid which should be contained in a wooden vessel. The liquid consists of one gallon of soft warm water, one-half ounce of commercial sulphuric acid, and one-third of a pound of corn starch. After compounding, this liquid should be stirred about ten minutes before the skin is introduced. When the pelt is put in, it should be pressed well under, and stirred with a wooden paddle for thirty minutes or until it begins to thicken and turn a dark reddish color. Then take it out and hang it up, and let it drain about thirty minutes, after which put it into a weak lye made either from wood ashes or from concentrated lye which may be bought in any grocery store; strain the liquid before putting the hide in. Stir the hide in this lye about thirty minutes again, or until you are satisfied the lye has had time to neutralize the acid from the preceding bath. The object of this latter bath is to counteract the acid effect of the former. Take out of the lye and hang up and let it drain about one hour, and then give it a thorough washing on a wash-board with plenty of soap and warm water. Rinse perfectly clean and again hang up to drain. When the hair is nearly dry, tack it again on the floor taking care to stretch it into proper shape. When partly dry take it up and rub and pull it until soft, which completes the tanning. We are then ready to place the skull in position. After skinning the head, remove the brains and eyeballs, and then boil the skull until all the flesh can be scraped from the bones. The skull being ready, fill the eye sockets with plaster paris made with water into the consistency of a plaster, and then set in a pair of suitable glass eyes, being careful to so adjust them as to give them the appearance of natural eyes.

The under jaw should be wired to the upper, or bound securely to it by a small cord or wire. Saw the skull and back corners of under jaw off, so that when the sawed surface is fastened down flat on a board, the nose will be somewhat elevated from the floor, while the back part of the skull will lie on the floor. Fasten the skull to the small boards with wire, and then stretch the skin over it. Soak only the skin of the head in water, and sew up the mouth by drawing the lower lips up under the upper in the natural way. The skin should be inside out while the lips are being sewed. If in the process of skinning you have cut the skin clear out through the lip, it must be sewed together far enough to inclose the skull and the under board, which should not extend more than two inches back of the skull. Build out the end of the nose with plaster paris to make it natural shape and draw the skin over the skull. Be very careful to have the nose and eyes all right. Tack the skin to the under side of the board which should in width correspond to the thickness of the dog's neck. The edges of the neck skin should be sutured together under the board and tacked to it. Cut the skin, which is lapped under, following the line of back end of board out to the side edges of the board. Turn this flap out and sew the underlying end to the upper skin, or to the skin of back of neck, and close to the back end of board; this brings the skin on a level of under side of board; now round off the turned out flap so as to conform with the line of the neck and fore legs; the head will now need some stuffing to round it up in proper shape. Take some cotton and put it in through the ears with a wire, placing it where needed to give the head a proper shape; arrange the skin around the eyes, nose and mouth; then let dry before lining. Put a good supply of the alcohol and corrosive sublimate on the ears, nose and lips, as soon as head is dressed. Now for the lining, procure a sufficient amount of felt to line it, say of a dark green color; then get enough to make a strip one and a half inches wide and long enough to go around the edge of hide; this should be an orange color.

Take this strip and pink one edge of it, then baste the lining on to skin letting the edges of it project about two inches out from edge of hide. Now comb the hair around the edges of hide back towards the center of same. Then take the pinked strip and lay it on the hair side of skin, the smooth edge along the edge of the hide and the pinked edge laying back on the hair; then sew the pinked strip, lining and hide together, running close to the edge of hide. Now turn the pinked strip out on to the projecting lining and sew another seam around just out side of the hide, thus sewing the lining and the pinked strip together. Now pink the out edge of lining to suit your taste; then take some yarn or zephyr and knot the lining and hide together, the same as in knotting a comfort, and the job is complete.

To make a robe sew a number of hides together and line them, simply binding the edge with a straight strip.

TO SOFTEN HIDES.

Where a hide has become hard by getting wet, or from some other cause after tanning, wet it on flesh side with water and hang up. When partly dry give it a thorough rubbing. If that fails to soften it, oil it with Neatsfoot oil and rub it in well.

TO CLEAN RUGS AND FURS.

When rugs and furs become dirty from use take hardwood sawdust, dampen it and rub well into hide. Shake out and repeat until clean.

TO KILL MOTHS IN FURS, ETC.

Sprinkle them well with arsenic shaking it well down on hide. Tie up and lay away for a few days. Then dust all the arsenic out and clean with the sawdust. In stuffed birds use the arsenic alone, but handle it with care for it is a deadly poison.

EXTRAS.

A FAMILY LINIMENT.

Alcohol	1 pint
Gum Camphor	½ ounce
Aqua ammonia	3 ounces
Oil of Sassafras	½ ounce
Laudanum	½ ounce

Mix.

Wishing to use the above for neuralgia or rheumatism, add to three ounces of the preparation 4 grains menthol (in crystals).

FIVE DAY CORN OR WART CURE.

Cannabis	5 grains
Salicylic Acid	29 grains
Collodion	½ ounce
Caster Oil	10 drops

Directions:

Mix thoroughly, and before applying dampen the corn or wart with turpentine; then apply, being careful not to get the preparation on the sound flesh. Repeat this every night before going to bed for five successive nights. Now bind a slice of lemon on excrescence, let it remain one hour, and then wash and scrape off corn or wart. You may of course repeat as often as necessary, or until excrescence is entirely removed.

FURNITURE POLISH.

Turpentine	½ pint
Yellow lubricating oil	½ pint
Muriatic acid	1 ounce

Directions:

Mix and let stand two days, and then apply with a sponge after which thoroughly dry with a woolen cloth.

WHITE CEMENT.

| Best white glue | ½ pound |
| Soft water | 2½ pints |

Directions:

Mix and heat over a slow fire until dissolved; then stir in 4 ounces dry white lead, ¼ pint alcohol, and 1 ounce aqua ammonia.

Lector House believes that a society develops through a two-fold approach of continuous learning and adaptation, which is derived from the study of classic literary works spread across the historic timeline of literature records. Therefore, we aim at reviving, repairing and redeveloping all those inaccessible or damaged but historically as well as culturally important literature across subjects so that the future generations may have an opportunity to study and learn from past works to embark upon a journey of creating a better future.

This book is a result of an effort made by Lector House towards making a contribution to the preservation and repair of original ancient works which might hold historical significance to the approach of continuous learning across subjects.

HAPPY READING & LEARNING!

LECTOR HOUSE LLP
E-MAIL: lectorpublishing@gmail.com